THINKING ON EARTHQUAKES IN EARLY MODERN EUROPE

This book is the first extensive study of ideas on earthquakes before the Lisbon earthquake in 1755. The earthquake had a deep impact on European culture, and the reactions to it stood in a long tradition that, before this study, had yet to be explored in detail.

Thinking on Earthquakes investigates both scholarly theories and views that were propagated among the early modern European population. Through a chronological approach, Vermij reveals that in contrast to the Ancient and medieval philosophers who suggested rational explanations for earthquakes, supernatural ideas made a powerful comeback in the sixteenth century. By analysing a variety of sources such as pamphlets, sermons, and treatises, this study shows how changes in the ideas on earthquakes were a result of social and political demands as well as from improvements in the means of communication, rather than from scientific methods. Thus, Vermij presents an illuminating case for the production of knowledge in early modern Europe.

A range of events are explored, including the Ferrara earthquake in 1570 and the Vienna earthquake in 1590, making this study an invaluable source for students and scholars of the history of science and the history of ideas in early modern Europe.

Rienk Vermij obtained his PhD in 1991 at Utrecht University, the Netherlands, and is presently Professor at the Department of History of Science of the University of Oklahoma. He has published on several aspects of early modern intellectual culture. Among his books is *The Calvinist Copernicans. The Reception of the New Astronomy in the Dutch Republic 1575–1750* (2002).

THINKING ON EARTHQUAKES IN EARLY MODERN EUROPE

Firm Beliefs on Shaky Ground

Rienk Vermij

Routledge
Taylor & Francis Group

LONDON AND NEW YORK

First published 2021
by Routledge
2 Park Square, Milton Park, Abingdon, Oxon OX14 4RN

and by Routledge
52 Vanderbilt Avenue, New York, NY 10017

Routledge is an imprint of the Taylor & Francis Group, an informa business

British Library Cataloguing-in-Publication Data
A catalogue record for this book is available from the British Library

Library of Congress Cataloging-in-Publication Data
Names: Vermij, Rienk, author.
Title: Thinking on earthquakes in early modern Europe: firm beliefs on
shaky ground / Rienk Vermij.
Description: Milton Park, Abingdon, Oxon; New York: Routledge, 2021. |
Includes bibliographical references and index.
Identifiers: LCCN 2020028663 | ISBN 9780367492182 (paperback) |
ISBN 9780367492199 (hardback) | ISBN 9781003045083 (ebook)
Subjects: LCSH: Earthquakes—Europe—History—16th century. |
Earthquakes—Europe—History—17th century. | Natural disasters—
Europe—History. | Earthquakes—Social aspects. | Civilization.
Classification: LCC QE536.2.E85 V47 2021 |
DDC 511.22094/09031—dc23
LC record available at https://lccn.loc.gov/2020028663

ISBN: 978-0-367-49219-9 (hbk)
ISBN: 978-0-367-49218-2 (pbk)
ISBN: 978-1-003-04508-3 (ebk)

Typeset in Bembo
by codeMantra

CONTENTS

PREFACE

This book is the result of research that spans several decades. It all started in 1997 with a fellowship at the Herzog August Library at Wolfenbüttel, Germany, which offered me excellent research opportunities. At the time, this resulted in several articles, but far from exhausted the subject. The project kept me occupied, but because of many other projects and obligations, it proceeded only slowly. Fortunately, my present employer, the University of Oklahoma, allotted me generous research time, including a summer stipend. The Descartes Center at Utrecht offered a welcoming environment during my sabbatical in 2013–2014. I owe gratitude to these institutions and apologies if they had to wait a bit long for the final result.

The long history of this project had some unforeseen consequences. When I started many years ago, digitalization was in its infancy. Because many sources were rare and often hidden in far-away places, studying them was expensive and demanded extensive travel. Since then, digitalization has made many works better accessible. This made my research much easier and has allowed me to fill in several gaps in my narrative. On the other hand, some of my earlier choices in selecting my material suddenly seemed somewhat less fortunate. That made for some extra work, although I hope I can be forgiven that I did not start all over from scratch. The overview is certainly not all-comprehensive. There is hardly a phenomenon on which has been written more than on earthquakes, and even today, many sources are still unavailable or hard to come by. At some point, one has to decide that what one has is good enough. Still more material might have resulted in a bigger, but not necessarily a better book.

In spite of all the digitalization, in many cases, I was dependent upon paper copies. I owe a great debt to the staff of the many libraries where I consulted materials in the course of the years. This is particularly true for the staff of the Herzog August Library at Wolfenbüttel and of the History of Science Collections at the University of Oklahoma. Also, I owe gratitude to the city archive of Schaffhausen and the central library of the canton Lucerne for obligingly sending me materials.

An embarrassing consequence of the long time the project took me is that I have lost track of many persons who in those years have been of assistance. As I discovered during my research, the history of earthquakes fascinates many people and many colleagues took an interest in the project. They often spontaneously sent or offered information, pointed to literature, suggested sources, or helped me out with unfamiliar languages. It is impossible to mention or even to recall everybody who in some way or other offered assistance or encouragement, and it is therefore with some uneasiness, and with sincere apologies to everybody I overlooked, that I recall the following names. A special thanks should go to my colleague at Oklahoma, emeritus professor Ken Taylor, who read and commented on the entire manuscript. Others whom I remember are, in alphabetical order, Ann Blair, Kerrewin van Blanken, Jan Willem Buisman, Patrizia Conforti, Bruno Figliuolo, Georges Maurice Frechet, Willem Frijhoff, Manfred Jakubowski-Tiessen, Eric Jorink, Maija Kallinen, Hartmut Lehmann, Ulrich Löffler, Kerry Magruder, Dieter Merzbacher, Cornelia Niekus-Moore, Augusto Placanica, Jo Spaans, and Joachim von der Thüsen. I am also happy for the support of my colleagues at the Department of History of Science at Norman, Oklahoma, and of the Descartes Center at Utrecht University, as well as for the stimulating discussions with the many fellows at Wolfenbüttel during my stay in 1997 and at several later occasions. I am responsible for the final result, but I would not have gotten there without everybody's support.

Translations into English are mine, unless another source is indicated. For biblical quotes, I used the King James Version as being closest to the period.

Part of this research was made possible by a generous grant from the Herzog August Library, Wolfenbüttel, Germany.

INTRODUCTION

This is a book on the history of science in Europe in the critical period that is traditionally referred to as the "scientific revolution". Although a number of familiar names will recur in these pages, the book focusses on a less familiar topic within that period: the science of earthquakes. Of all natural phenomena, earthquakes may well be those on which earlier authors have spilled most ink. They were a concern of philosophers and scientists, but also of theologians, ministers, and historians, and basically of anybody who experienced them.

Not only were earthquakes a topic of intense scrutiny. They are also a subject where scientific and religious interpretations interfered and sometimes clashed. Earthquakes, because of their extraordinary nature, the awe and surprise they almost invariably inspire, and the often very real danger they present have always been linked to the supernatural. For most of the early modern authors on earthquakes, the religious interpretation was an important question. Their writings refer to scientific theories, but also to religious sensibilities and political goals. Changing interpretations of earthquakes did reflect changing attitudes with respect to nature and reality in general. In this way, earthquakes offer insights into the complicated relationships between learned culture and popular culture and between religious ideas and philosophical explanations.

The reactions to the earthquake of Lisbon of 1755 have long been recognized as an important moment in European cultural history and have been the subject of extensive research.[1] It is therefore somewhat striking that the many other earthquakes have been mostly ignored. Ideas on earthquakes have hardly been studied as a part of early modern science or culture, although there are a few overviews of past theories.[2] Still, it was an established field. Earthquakes were part of the study of "meteorology" – not meteorology in the modern sense, but works in the tradition of Aristotle's *Meteorologia*.

There are several reasons for this neglect. Most importantly, until fairly recently historians of science have mostly been interested in discoveries and innovations. In early modern seismology, these were hard to find. It was only in the nineteenth century that geologists realized that earthquakes occurred along fault lines, and an understanding of underlying mechanisms had to wait until the second half of the twentieth century. No such breakthroughs happened in the early modern period. In 1500, people had no idea what caused earthquakes and in 1800, this was not different.

What is true of earthquakes is true for the field of early modern meteorology in general: on the theoretical level, it hardly saw any "progress". The few exceptions were quickly taken out of the field of meteorology – comets came to be studied as part of astronomy and cosmology, rainbows as part of optics. The field of meteorology itself remained largely unaffected by the intellectual changes in the seventeenth century and consequently is largely absent from the standard histories of the "scientific revolution".[3]

Even now that historians of science have abandoned the idea that historical developments should be analysed not within a modern framework of the history of human progress, but within the intellectual and institutional framework of the period itself, the traditional topics are still dominant. Historians ask new questions to familiar material, but still find it difficult to completely ignore the familiar demarcations. If we take our task seriously of understanding the development of learning and the sciences in the early modern period, we cannot ignore a field like earthquakes. Early modern ideas on earthquakes are a mere detail in the history of scientific theory as traditionally conceived, but they throw important light on the relation of the sciences to the social, political, and ecclesiastical structures in the early modern period.

So far, the history of earthquakes in the early modern period has been mostly studied from two different perspectives. One is the perspective of geoscience. Geologists have taken great pains to recover evidence of past earthquakes, with the purpose of finding clues of the geological processes that have played and are playing over time. To this purpose, they have unearthed many historical sources and documented many events, which makes their work valuable for general historians as well. However, in the end they have studied these sources for what they say about geological processes, not human history.[4]

The second perspective is the relatively new field of disaster studies. In recent years, the impact on humans and societies of major disasters, both in the more recent and the more distant past, has grown into a field of intense scholarly interest. From this perspective, the effects of earthquakes have been studied as well; that is to say, major earthquakes, since definitely not all earthquakes are disasters. Historians of disasters are typically not interested in smaller earthquakes, still the vast majority. Moreover, most of them have more affinity with the history of mentalities than with the history of ideas: they focus on social and psychological effects, not on scientific or philosophical thought. Apart from the events themselves, they are also interested in the long-term effects and questions of resilience and recovery.[5]

These two fields have documented many events and brought valuable insights, but obviously have their limitations, in particular when it comes to history of science. In this book, ideas on earthquakes are studied as a case of the wider intellectual and scientific changes of the period. Even though there was no "progress" in the modern sense, the field was not completely static. So, the question is what made people frame their ideas on earthquakes in a situation wherein empirical reality, in the form of mathematical or empirical data, offered little or no clues. The fact that the theories of earthquakes changed without any mathematical or experimental evidence, and without the new theories being in any way empirically more adequate than the old ones, allows us to study the formation of theories without our view being troubled by notions of "progress". Earthquakes are therefore ideally suited for a historical case study of knowledge production. In the end, this book is not so much a history of earthquakes, as of knowledge.

This approach is of course well in line with the more recent tendencies in the history of early modern science, where intellectual changes are now studied within the larger religious, political, technological, and institutional contexts. Particularly relevant to this book is the work that has been done on the development of the sciences in the context of religious institutions. The place and function of the various disciplines at the newly confessionalized universities and within religious orders has been the subject of intense study in recent decades. In particular, important studies have been written on the developments in natural philosophy at the University of Wittenberg, which put natural philosophy into the service of Lutheranism, and within the Jesuit order, which aimed at creating a form of natural knowledge that was in accordance with Roman Catholic orthodoxy. Both the Jesuits and the University of Wittenberg were extremely influential and managed to impose their views more widely in Europe.[6]

It should be admitted that such studies in many cases mostly focussed on the very fields that have traditionally drawn most interest from historians of the scientific revolution, such as astronomy and cosmology, anatomy, and the principles of natural philosophy. However, some less traditional fields have been studied as well, as in the case of the Lutherans their teaching and promotion of astrology. The history of meteorology too, although still a largely ignored subject, has recently been the subject of a few studies in this context.[7] The present work is a contribution to this growing field of studies for, as will become clear, in the theorizing about earthquakes as well, both the Jesuits and the scholars at Wittenberg were very influential.

Nevertheless, my book takes a different approach from these earlier studies in at least three different ways. Earlier studies typically focus on either a Protestant denomination, or on the Catholic Church, or on a particular congregation like the Jesuits. I feel it is important to emphasize the basic unity of the period. Consequently, I will deal with religious culture as a whole. This creates a problem for labelling. Although scholars have recognized that the intellectual activity in Reformation and Counter Reformation culture answers to certain common characteristics and should be regarded a distinct movement, so far there exists no generally accepted common denominator that on the one hand encompasses both "Jesuit science", "science at Wittenberg", and so on, and on the other hand sets it apart from forms of scholarship that were not theologically motivated or embedded in the educational or ecclesiastic structures of the time. Older historiography often made with terms like "religious", which is misleading, since we are talking of a very time-specific and theologically peculiar form of religious thinking, distinct from medieval or modern conceptions. Terms like "orthodox" or "theologically-inspired" have similar objections. The term "baroque" is sometimes used for the period, but this term has all kinds of other connotations, and moreover has already been appropriated in history of science to some other uses. For lack of a better term, I have decided to use in this book the form "confessionalized science" or "confessionalized philosophy" whenever I need a general term. This is inspired by the German term *Konfessionelles Zeitalter* (confessional era) for the period of the Reformation and the wars of religion, and justified by the dominant role that the individual Churches, each with their own confession, came to play in social organization. So, the wider subject of this book is the confessionalization (both by Protestants and Catholics) and deconfessionalization of natural philosophy in the early modern period.

In the second place, my book does not limit itself to a discussion of ideas as they were formulated as part of some philosophical programme, but tries to connect them to their wider social and religious use. For most of the time, the causes and nature of earthquakes may have been the subject of abstract academic speculation; however, once a real earthquake occurred, they were everybody's concern. Theologians and philosophers were required to explain the phenomenon to the people in a way that was in line with religious and social demands and much of their theorizing actually had this goal in mind. Consequently, the book will not only discuss academic textbooks and philosophical treatises, but also the pamphlets, sermons, and other works in the vernacular that appeared to instruct the people after some major or minor earthquakes had caused a sensation.

In the third place, I will not only discuss the beginnings of "confessionalized science" and its growth and development, but will also pose the question of its demise, or at least how it lost its dominant position. Traditional history of science has always been preoccupied with discoveries and inventions. The abandonment of old ideas was simply seen as the natural result of the emergence of new and better theories. It is only very recently that historians have come to realize that the situation is more complicated and that the marginalization of fields like astrology or alchemy constitutes an important element of the "scientific revolution" in its own right. Likewise, the developments within confessionalized science are part of the larger intellectual developments of the seventeenth century. In order to understand the emergence of new ideas, one should know what happened to the old.

The book basically has the form of a chronological narrative. It is of course informed by many points debated in modern scholarship, but in the text, these are not always made explicit. Some readers may feel this unhelpful, but I do not apologize. The history of earthquakes is largely unexplored territory and many of the sources are not easily accessible. I therefore felt it important to give first of all a fairly comprehensive overview and discuss a number of authors in greater or lesser detail. Including elaborate historiographical discussions would alter the character of the text and require way too much space.

The organization of the text, then, is as follows. In the first part, I lay out some necessary groundwork. In order to understand the changing views on earthquakes, we have to be familiar with what was known about them at the beginning of our period. That concerns first of all the existing scholarly traditions on the subject. I will therefore give a brief summary of the ideas that had been transmitted on earthquakes since antiquity, and the way these had obtained shape during the Middle Ages. However, I will not discuss the Middle Ages in any depth, since this period, before the invention of printing, puts very different demands on the researcher. Nor will I discuss the non-European world. This book is a study of knowledge formation in Europe in the early modern period, not an encyclopaedia.

Apart from medieval scholasticism, Renaissance humanism was a powerful intellectual force in the era of the Reformation and we cannot well understand developments in the sixteenth and seventeenth centuries without paying attention to the influence of humanist scholarship. It will be necessary to include a chapter on humanist views on earthquakes before entering upon the views of Protestant and Catholic reformers.

However, apart from scholarly traditions, we also have to deal with reports of events that happened at the time itself. Earthquakes are very real to those who live through them and knowledge would be based on actual experience as well. On the other hand, earthquakes are also relatively rare and to most people of Europe, first-hand experience was limited. However, news about and descriptions of earthquakes became more widely available in the early modern period as a result of the growth of the means of communication. This certainly is one of the main driving forces behind the changing views of earthquakes, but its effects were not unambiguous. People could use the available information both for understanding God's judgements, and for purposes of natural history.

The second part is the main part of the book. It deals with the ideas and traditions as they developed in the sixteenth and seventeenth century under the influence of Reformation and Counter-Reformation. For the theologians and philosophers of the period, earthquakes and other strange phenomena were manifestations of God's action in the world. They heavily influenced the later interpretation of these events. First of all, we will discuss the theories that were put forward in university textbooks and similar works. The Lutherans at the University of Wittenberg were at the forefront, as were the Jesuits on the Catholic side, but other Protestant denominations and religious orders played their parts as well. Although these scholars did remain within the Aristotelian tradition of meteorology that had been dominant in the Middle Ages, other elements were introduced as well, notably an interpretation of earthquakes as portents (for which they could refer to Pliny and other classical authors) and a growing emphasis on the Bible. Whereas medieval scholastics and Renaissance humanists had on the whole preferred a philosophical, largely secular understanding of earthquakes, sixteenth and seventeenth-century scholars emphasized their religious meaning. At the same time, they were very wary about magic and what they considered superstition. By imposing their religious orthodoxy, they may have de-secularized philosophy, but at the same time, they to a certain extent "disenchanted" wider society.

Reformation or Counter-Reformation science had as their goal the transformation of people's ideas and behaviour and should be studied in relation to their social goals and effects. I therefore am discussing not just the scholarly traditions as they were taught at universities, in the form of textbooks or disputations, or discussed among the learned, but also the way that ideas on earthquakes were put to use by the clergy in addressing the common populace, in the form of pamphlets, sermons, and histories aimed at a wider audience. As will be shown, ecclesiastical teaching did transform the debate on earthquakes. Ideally, learned and popular approaches had to be in agreement, but not all authors succeeded in fully accomplishing that. Still, the religious and edifying requirements set a definite mark on the theories as they were defended by the learned, so that one is justified to speak of a distinct form of scholarship in this period, based on confessional thinking.

We cannot study the history of the "confessionalized science" of the sixteenth and seventeenth centuries without posing, finally, the question how by the end of the seventeenth century it lost its intellectual appeal and its dominant position in academic teaching and in society. The third part of the book therefore is devoted to the emergence of the new physical outlook of the seventeenth century. There will be some discussion of the first half of the eighteenth century as well, but only up to the year 1755. The earthquake of Lisbon, which happened in that year, and its various effects have been the subject of extensive research, which makes a treatment at this place superfluous.

A full discussion of the reasons why the confessional approach lost appeal falls beyond the scope of this book. Just as the rise of this approach was made possible by the social, political, and religious upheavals of the Reformation era, so its gradual decline too was initiated by changes in the wider socio-political world. Most of these fall beyond our purview, but there are several points we need to consider. In the first place, in a history of the sciences, we obviously have to be aware of the establishment of the "new philosophy" and the new standards of natural research and natural explanation. Even though the new sciences hardly contributed to a better understanding of earthquakes in a direct way, one might still assume that the new intellectual framework would impact methods and theory. There is certainly a connection here, but the direct impact of the new standards appears limited. In the second place, the changing religious landscape and changing religious sensibilities of the period might appear as a plausible cause of changes. This is certainly an important point, but it should not be overstated.

A more detailed study of the reactions to earthquakes will show that the shift in religious sensibilities was actually not so great as might appear when only looking at some isolated topics. The third and probably most important factor to be discussed are the means of information and communication, which had considerably advanced since the early days of the Reformation and which set new standards for reports on all kinds of events. The new empirical outlook of the "scientific revolution" may actually owe more to the shifting standards of communication among merchants, sailors, administrators, and others, than to any philosophical or scholarly programme.

Notes

1 E.g. Löffler (1999); Buisman (1992) 79–107. A full overview of the literature goes beyond the scope of these pages. Kendrick (1956) is an old, but still useful overview.
2 E.g. Oeser (1992); Taylor (1975).
3 For overviews of early modern meteorology, see Hellmann (1917); Heniger (1960); Gilson (1921). Only in recent years historians have started to take the topic more seriously, see Martin (2011); Borrelli (2008); Vermij (2010); Céard (2013).
4 See for instance the articles in Stucchi ed. (1993); Fréchet, Meghraoui and Stucchi (2008). See also Neilson a.o. (1984).
5 For instance Pfister (2002); Bogucka (1999); Jakubowski-Tiessen and Lehmann (2003); Jakubowski-Tiessen (1992); Berlioz (1998); Delumeau and Lequin (1987).
6 E.g. Kusukawa (1995); Methuen (1996); Brosseder (2004); Jorink (2010); Feingold (2003); Remmert (2007); Ashworth (1990).
7 See Vermij (2010) on Wittenberg and Martin (2011), who gives a prominent role to the Jesuits, in particular Niccolò Cabeo.

PART I

Scientific, philosophical, and religious traditions up to the Renaissance

1

EXPERIENCING EARTHQUAKES

This book is a history of ideas, not of mentalities. My focus is on the development of learned theory. However, that is not to say that I can completely ignore other approaches. Ideally, scholars may have developed their ideas in the splendid isolation of their studies, leafing through centuries-old texts. In practice, they were as much part of society as anybody else. If an earthquake struck, they shared in the common experience. It is unlikely that this would not have any repercussions on their thought. We may well start, therefore, by paying some attention to these common experiences.

Not all earthquake experiences are the same. In some countries, earthquakes are well-known phenomena. When an earthquake happens, people will immediately be on the alert, well understanding the threat; but if the quake turns out to be a minor one, they will quickly move on with their lives. In other countries, earthquakes are very rare events. Many people will never have lived through one. In these regions, even relatively small earthquakes are quite upsetting.

People unfamiliar with earthquakes often had difficulty realizing what was going on. The English scholar Gabriel Harvey related how the earthquake of 1580 happened while he was in a company playing at cards. Since people were uncertain of what had happened, he asked his host to send a servant into town to inquire whether the movement had been general. When this proved to be the case, he decided it must have been an earthquake.[1] In that sense, earthquakes can be called social constructs.

In recent times, people sometimes initially suspect a heavy truck or other machines. Similar interpretations occurred in the past. "I doubted the fall of a large Piece of Timber, or Stone-Work", wrote the reverend Paschall to Robert Hooke about an earthquake in January 1680.[2] The Lutheran minister Tobias Wagner wrote on an earthquake in 1655 at Württemberg, which woke him up from sleep: "I had the idea that a bag filled with flour fell down from the upper part of my room with great violence, as such a fall is not uncommon with me, and happens more often at night."[3] When the mathematician Christiaan Huygens sensed the earthquake of November 1692 at his home near The Hague, he initially related it to the war in the southern Netherlands and surmised the explosion of a gunpowder magazine at Duinkerke, some 160 kilometres away.[4]

Huygens' brother Constantijn, who as secretary to King William III of England took part in the military campaign, was much closer to the epicentre. When the earthquake occurred, he and others were just having dinner in a tent. Seeing the table shaking, Constantijn's first thought was, as he wrote in his diary, that some horse had got entangled in the strings. Once people realized what was happening, everybody rose and ran away, shouting "an earthquake, an earthquake". In the house where the King was staying, there was a thronging of people trying to escape. The King himself had fallen in the hustle. One other lord, Montpouillon, had injured his shin. In the end, the panic may have caused more harm than the earthquake itself.[5]

In countries where earthquakes are more frequent, like Italy, people had no trouble realizing what was going on and their reactions would be more adequate. The Italian author Agatio Di Somma experienced the great Calabrian earthquake of 1638 in his hometown, Catanzaro. With many other people, he happened to be in the city square when the earthquake struck. "As soon as the unforeseen buzzing in the air, of which I spoke, made itself felt, everybody, murmuring: 'terremoto, terremoto', as if they had wings, fled to the widest part of the square, as the least dangerous."[6]

Fear and panic

Serious earthquakes could be extremely traumatic experiences. The clearest attestation of the stress they caused to the population is given by the spontaneous outbursts of panic that often occurred after an earthquake. Rumours about strange, maybe supernatural appearances were eagerly told and believed. After the earthquake of 1590, people in Prague saw strange things at night: spooky people carrying lights, and a man in red who behaved strangely and then mysteriously disappeared. After the earthquake of Calabria of 1638, people reported on bloody and fiery crosses seen in the sky.[7]

Often, rumours circulated that the earthquake had been announced or prophesized.[8] An hour before the great earthquake of 1674 of Ambon in the East Indies, a Christian slave had seen in the hall of the fortress a person standing on the pulpit with white face and hands, a book in his one hand and a candle in the other, as if reading, with long black clothes and on his head a crown of thorns. The naturalist Rumphius, who reported this, was somewhat sceptical ("Whatever the truth may be we leave it to the Reader's judgment") but could not deny that the slave who saw the vision was a credible person, who gave his report repeatedly to several respectable persons "and he never varied his tale".[9] In Lima, after the earthquake of 1746, it was told that 21 days before the disaster struck, an honourable man had seen in a vision that all the houses of Callao, the harbour of Lima, were on fire. Moreover, less than a month before the quake, a pious nun had explained to her confessor that the anger of God was against the city and its inhabitants, but that she would die before she could witness the effects of the divine justice. Indeed, she died 13 days before the earthquake.[10]

However, such alleged prophecies not just concerned the past, but the future as well. After the great earthquake of Vienna of 1590, the Fugger newsletter reported rumours that some anonymous scholars had predicted that a still much larger earthquake would follow, or even that Vienna would be completely destroyed in four weeks.[11] The protocols of a Franciscan convent in Innsbruck in Austria report on an earthquake in 1687:

> The confusion was increased by a rumour that was spread by the Jesuit father Ferdinand Orban, mathematician and on holidays court preacher in our church. He wrongly asserted that within 24 hours, the earth would quake still much more terribly than the first time. Thereupon everybody who had the opportunity fled from the city to the surrounding farmsteads or the public fields.[12]

Concern about aftershocks was of course understandable, but the fears among the population went far beyond that. After the Italian earthquake of January 1703, a panic arose at Rome in the middle of the night of February 3. People started banging on doors; a voice shouted that the Pope had pronounced that the whole city would perish that night at the tenth hour, and that everybody should flee. Indeed, many fled in confusion; according to the physician Giorgio Baglivi, who related the event, not even a third of the population stayed behind. Once they reached the open fields, the people implored God to save the city for the sake of the innocent children. When it was announced that the alarm was false, they returned home. Those responsible for the panic could not be traced, so that many suspected that it was the work of demons. They pointed out that that same night at the same hour, a voice threatening imminent destruction was heard in villages and suburbs at many miles distant.[13]

A similar panic is reported from Lima in 1605, after rumours of an earthquake in the south had reached the city and a Franciscan friar had warned that Lima, because of its many sins, might be destroyed likewise. Churches and convents opened, lit up candles and displayed the holy sacrament. People pressed priests in the streets to be confessed. Some flogged themselves, others gave alms, and some hastily married to sanction an illegitimate relationship.[14] Panic struck again in Lima after the big earthquake of 1746; many fled to the hills when a rumour spread that the sea would flood the city.[15] (Lima is actually several miles inland and had shown to be out of reach of tsunamis.) For a decade after 1746, the population of Lima would be tormented by premonitions of impending doom.[16] Even in full Enlightenment London, in 1750, two minor quakes exactly one month apart caused such concern that many notables fled the town when a prophecy was circulated that after yet another month the city would be completely destroyed by a third earthquake.[17]

Prognostications that earlier had drawn little attention now suddenly might appear ominous. A pamphlet printed at Nuremberg in 1627 announces in the title a report of the earthquake that hit the region of Apulia (Italy) with a short description of the kingdom of Naples. In reality, it offers a long description of the kingdom of Naples, but has little to say on the earthquake. The reason to publish the pamphlet in spite of this lack of relevant information must lay in the appendix, also announced in the title: a long extract from a prognostication by David Herlicius (or Herlitz), a well-known mathematician and physician. In this prognostication (for 1627), he had foretold devastating earthquakes. Herlicius explained that these disasters would strike because of our sins: "Such terrestrial signs will come to us, because we will not convert or improve because of the celestial ones".[18] The edifying message, driven home by a fulfilled prophecy, must have been the reason to publish this in other respects rather uninformative pamphlet in the first place.

The devastating earthquake of Calabria of 1638 gave rise to frightening predictions. A pamphlet published in the German city of Breslau (present-day Wrocław, in Poland) prints various reports of this earthquake. One of these is a letter from Prague, dated May 16, 1638. "From here, there is no other news but that there is great fear and worry among part of both the notables and the common people, because of the prophecy of the astronomer who is confined in Venice." This (otherwise-unknown) astronomer had predicted that 19 July, when the sun would enter the sign of the Lion, the countries and cities under that sign would be hit by disaster: darkness, storms, airy voices. This would be followed, 23 July between 7 and 8 o'clock in the evening, by a severe earthquake.

> This astronomer appears to have prophesied about the island [sic] Calabria as well, where 9 cities and 209 villages have been destroyed. Therefore, people here are very worried that harm may befall us, because he specifically mentioned Prague. You may believe me that there is such a fear among part of the population here, that many pack their best belongings and get away. Wherever there are two persons standing together in the streets, they talk about nothing but this prophecy and the destruction of the city of Prague.[19]

In Calabria itself, fear was still more intense. A rumour emerged that the quake had been foretold, apparently in some prognostication, by the physician and astrologer Pietro Paolo Sassonio from Cosenza in northern Calabria. His apparent success made Sassonio push forward and predict another earthquake for May 5, which would dwarf the earlier one. The rumour spread like wildfire. People now suddenly remembered still older predictions, among others those by the famous twelfth-century visionary writer Joachim of Fiore. Reports of weeping statues of St. Mary confirmed their fears. As the unrest grew, Sassonio was summoned before the Royal tribunal at Naples and sent to the galleys as a perturbator of the peace. (In the end, the inquisition confined him to a monastery in Capri.) This did not undo the popular anxiety, which was kept alive by frequent aftershocks. On 4 and 5 May, the days mentioned in the prediction, people pretended to be at ease, but were actually quite tense. According to Di Somma, who described the whole episode at length, many went for a walk, casually as it seemed, but they made sure to stay away from mountains or high buildings and watched every breath of wind. When at the end of the day nothing had happened, they took heart again and laughed at their fear. People started feeling secure again, until another minor quake on June 8 put everybody back on edge.[20]

One might assume that such stories were included in the records for literary effect, as these reports (as will be discussed later) were often written for edifying purposes. Indeed, it was not rare that some over-zealous priests stoked rumours in order to convert the people by fear. On the whole, however, such predictions were not regarded as edifying, but rather as dangerous. Authorities had good reason to be wary of them, for once they were out, people interpreted them their own way, outside of anybody's control. This could cause serious problems for public order. Authorities did not encourage fear-mongering by almanac writers or others, as the case of Sassonio makes clear.

For this reason, most authors tended to downplay the significance of the various signs and predictions rather than to highlight them. Di Somma was quite critical of the wonders accompanying the Calabrian earthquake and attributed them to natural causes. In his case, edifying the people came down to poking fun at popular superstition. The story about the apparitions at Vienna was not reported in any edifying treatise, but in the newsletter of the banking house of Fugger. Whatever the reason for including them in the reports, it cannot reasonably be doubted that in most cases the stories went back to actual rumours.

Scientists had hardly a reason to be less afraid of earthquakes than other people, and until very recently, scientific explanations of earthquakes were hardly any better than theological ones. Still, it is also obvious that fears, as they emerge in panic or stories about apparitions and the like, were more prevalent in one period than in another. Even though natural, fear is also a result of the general outlook and expectations of the time. This was hardly dependent on the theoretical interpretations available, but rather on more general characteristics. The early modern period was a period of great uncertainty and insecurity in Europe, and this affected people's general view of the world.[21]

Piety

An obvious reaction to the frightening experience of an earthquake was to seek refuge in the rituals and explanations of the Church. In order to make sense of the event, people would have recourse to theological models: an earthquake was God's punishment for our sins, or it was an announcement of the Last Judgement. An anonymous witness of the earthquake in Scarperia (Italy) wrote afterwards: "I definitely believe that I have seen and experienced an image of the Last Day".[22] People would fall on their knees, pray, call on their patron saints, or confess their sins. The Anglican minister of Port Royal on Jamaica described how immediately after the devastating earthquake of 1692, the people "cry'd out to me to come and Pray with them". He made them kneel down and make a large ring, and

> I prayed with them near an hour ... the Earth working all the while with new motions, and tremblings, like the rowlings of the Sea; insomuch that sometimes when I was at Prayer I could hardly keep my self upon my knees.[23]

It is a widespread assumption that such a reaction is natural or in a sense primitive among people not trained in modern science. The evidence that people saw earthquakes in a religious framework is overwhelming indeed. However, we should realize that our sources are biased. Many authors on earthquakes were clergymen or people closely aligned with the established order. Whereas they only grudgingly included such wonder stories as discussed in the last section, they definitely seized upon the event to promote piety. Consequently, the edifying elements in their stories are highlighted. In some cases, one may ask whether one has not simply to do with pious inventions.

Ministers are also complaining about the lack of piety and spiritual renewal after an earthquake. Tobias Wagner published his sermons on a continuing string of earthquakes in 1655 with the argument that the world "for the most part gives little indication of true penitence for the horrible earthquake-warning".[24] From Port Royal came not just reports of prayers, but also complaints about pillaging, drinking, cursing, and whoring the night after the earthquake.[25] Many people appear to have found the chaos and devastation of an earthquake a good opportunity to enrich themselves in unlawful ways. Italian authors who suggested that the whole population was moved to penitence often at the same time praised the authorities for their tough measures against pillaging. Especially notorious was the pillaging after the 1693 Sicilian earthquake. Memorial inscriptions on buildings at Catania explicitly referred to it.[26]

That said, it is clear that religious formulas and rituals did play an important role. After all, these were readily available, or rather actively promoted, and many people must have interiorized them. One finds the religious interpretations not just in pamphlets and histories, but also in private letters and diaries. So, the Viennese noblewoman Eva Ungnadin wrote 18 September 1590 in a private letter:

> Not only have we had for some time all kinds of serious diseases and epidemics, we also have had four days on a row terrible big earthquakes, and still, which have caused much damage to churches and houses in the city, also in the countryside and to castles, so that we are in great danger. May

it please the almighty good God, for the sake of Christ, our only atoner and intermediary, gracefully to ward off from us His justified anger and the well-deserved punishment. I recommend us all to His protection....[27]

Outsiders might have a more sceptical view on the general piety. A German witness of the 1726 Palermo earthquake, apparently a soldier from the local garrison, admitted that he had been deeply shaken:

I and many others have been shaken in a way we will not forget in our lifetime. (…) He cannot be a reasonable human being, who, when to this point he had lived in sin, would not convert his heart and start leading a better life.[28]

Still, he was critical about the sudden wave of piety he saw around him. He pointed out that people were too busy praying to help save the victims. As to the real effects, he commented:

At the moment, now that they have the punishment right before their eyes and, following the predictions of some prophets (who have been arrested), fear the complete destruction of the city, everybody all of a sudden wants to convert. Whether and how they will keep their promises, and their conversion will last, God knows best, and time will tell.[29]

Fear is a natural reaction. The interpretation of earthquakes as divine punishments or as heralds of the dissolution of the world is not. Rather than a spontaneous reaction, it was the result of active Church teaching. Especially in the sixteenth and seventeenth centuries, for reasons which we will have to discuss, the various Churches often referred to disastrous or spectacular events like earthquakes when it came to instructing their flock. Their interpretation in terms of sin-and-punishment is not the result of an age-old, "primitive" state of mind, but a rather particular early modern way of understanding the world, varying with time and circumstances. We cannot simply talk about a religious view of earthquakes in general, but we will have to pay attention to shifting religious sensibilities and interpretations.

Curiosity

There is still another way people could react to earthquakes. Earthquakes were threatening, but also fascinating and sensational. They aroused fear, but also curiosity. Instead of praying, some people showed bravura. Renward Cysat, secretary of the city of Lucerne (Switzerland), wrote a detailed account of what happened in his hometown during the earthquake of 1601. The water in the river Reuss, which runs through the city as the outlet from Lake Lucerne, repeatedly withdrew into the lake, taking the ships with it, and then came back with force. According to Cysat, this phenomenon frightened the people more than the earthquake itself. Still, not everybody was just wringing their hands.

> The water had so far disappeared between the two parts of the city, that one could have crossed (so to speak) nearly with dry feet from the armoury to the mill, and several young people reportedly did so for a memorial.[30]

In some cases, it was a philosophical frame of mind that made people unwilling to be carried away by panic. Agatio Di Somma was at Catanzaro, socializing outdoors, when the city was struck by the great Calabrian earthquake of 1638. While the people were flying into the open, "I, who because of my travels was like a stranger in this city (still my hometown), and without experience in such events, followed the fleeing of the others much slower, and more observing and curious...."[31] Di Somma preferred to present himself as a detached, philosophically minded scholar.

Francesco Travagini reported on the occasion of the earthquake of 6 April 1667, which destroyed the Dalmatian city of Ragusa (present-day Dubrovnik):

> I was at that time in Venice, and happened to be resting in my bedroom (*in cubiculo*); in that way, the first impulse did not go unnoticed. When I realized this really was an earthquake, I immediately opened the window, and focused my attention so that nothing of what happened around me would escape me, especially those things which in an earlier earthquake had appeared to me as a noteworthy indication of an important truth.

Instead of being confused or panicked, or turning to prayer, Travagini realized that here was an opportunity for scientific observation.[32]

It is hard to read people's minds after such a long time. Whether they actually felt or behaved as they pretended is impossible to say. That is true both for people who assert that they saw an image of the Last Judgement and fell on their knees, and for people who claim that they were making scientific observations. In both cases, they present their behaviour according to a pre-existing model: in the one case Church doctrine, and in the other the ideal of the philosopher or scientist. In some cases, this modelling may have occurred after the fact, either consciously or as a trick their memory played on them. However, it can hardly be doubted that such models do have an effect on people's actual behaviour as well. There are plenty of reports of people spontaneously falling back on prayers or religious ceremonies during an earthquake. Likewise, the facts that occurred during or in the immediate aftermath of an earthquake could not have been found if people were too afraid take notice, or to care. The fact that Travagini was able to give a detailed description of the things that supposedly happened during the quake, demonstrates that he did take some notice of his surroundings, even if (as he himself admits) his observations were later supplemented by asking others what they had seen.[33] The question of how ideas on earthquakes developed to a large extent comes down to what mental models were available, models both of inter-pretation of the event itself and of how one was supposed to behave.

One cannot simply identify a detached view of things with a scientific out-look. Di Somma wrote that he was "observing and curious", but looking at his history as a whole, it appears that he was interested in the reactions of his fellow-humans, rather than in physical phenomena. His detachment was Stoic *apatheia* rather than scientific curiosity. This is different for Travagini, whose mind is set upon physical questions concerning the constitution of the world, and who shows no interest in classical philosophy. In other words, the rise of a scientific study of earthquakes has little to do with the rise of a detached, non-religious outlook, as such an outlook existed since antiquity. The question is why people at some point in time forgot about traditional philosophical attitudes and questions and instead focussed on natural phenomena.

So although it cannot be denied that over the course of many centuries, religious interpretations have become less dominant and scientific views have become more authoritative, the discussion cannot be reduced to an opposition between two poles, a religious and a secular one. Scientific thinkers first of all took issue with traditional philosophical approaches, including moral philos-ophy, more than with religious views. The theologians of the Reformation and Counter-Reformation were opposing older religious attitudes as much as they were combating naturalism. These developments were often interrelated. Theologians generally preferred a specific form of natural philosophy, whereas philosophers had a religious commitment. The resulting debate was a polyph-ony of voices wherein the development of the theme becomes clear only in course of time.

Notes

1 Harvey (1949/1966) 449–450. Harvey's account appears to have been stylized for literary reasons and may therefore not be completely accurate, but writing immedi-ately after the event, he had to remain close to the facts to remain credible.
2 Paschall to Hooke, printed in Derham ed. (1726) 54.
3 Wagner (1655) 74–75.
4 Huygens (1888–1950) XIX, 311.
5 Huygens (1877) 124–125. A very similar report of the events in the army camp was printed in *Unglücks-Chronica* (1692). d'Auvergne (1693) 61–62, claims that some soldiers of the garrison at Diksmuide initially thought that the French had under-mined the city and were going to blow them up.
6 Di Somma (1641) 62–63. "tutti, quasi che hauesser l'ali mormorando frà se Terremoto, Terremoto, sen volarono al più ampio sen della piazza, come meno pericoloso."
7 Fugger newsletter, printed in Gutdeutsch (1987) 114–115. Di Somma (1641) 59.
8 Cf. Schmidt (1994) 393.
9 Rumphius (1675/1997) 7, with translation on p. 53.
10 Walker (2008) 11, 44–45. Llano y Zapata (1748) 29–30. *True and particular relation* (1746) 75–182, 197–198. Cf. Frézier (1717) 365, on an alleged prophecy in Lima on the earthquake of 1682.
11 Text in Gutdeutsch et al. (1987) 116, 118.
12 Ruggenthaler (2016).

13 Baglivi (1704) 522–524. On fear because of astrological predictions on the occasion of this earthquake, see Grimaldi (1703) 16.
14 Osorio (2008) 23, referring to Diego de Ocaña, *Un viaje fascinante por la America Hispana del siglo XVI* (Madrid 1969), 101–103. Ocaña was a Spanish priest travelling in Peru at the time.
15 Llano y Zapata (1748) 2–3, 12. *True and particular account* (1746) 175–182.
16 Walker (2008) 22–51.
17 Rousseau (1969) 438–441; Kendrick (1956) 11–14. For a similar case in Palermo in 1726, see Ruffo (1727, German version) 29. Even as recent as December 1990, there was widespread concern because of a prediction of a second earthquake at New Madrid, Missouri, USA. MacDougall (2011) 118. See also 116–117 for contemporary reactions to the first New Madrid earthquakes in 1811–1812.
18 *Gründlicher Bericht* (1627) B2. On Herlicius, see Barnes (2016) 279–283.
19 *Einkommender Bericht* (1638) 14–15.
20 Di Somma (1641) 137–151.
21 Generally on the topic, Delumeau (1978).
22 *Erschröckliche Newe zeytung* (1542): "Ich glaub krefftig / das ich ein gleichnuess des Juengsten tages wol gesehen / unnd erfaren hab."
23 Heath (1692) 1.
24 Wagner (1655) dedication, see also p. 62 sub B.
25 Gragg (2009) 31.
26 Nicolosi (1982) 90–93, 309–311. *Sonderbahrer nachdenklicher bericht* (1695) 3.
27 Eva Ungnadin to Clusius, 1590 Sept 18. UB Leiden, VUL 101. Clusius project.
28 *Relation* (1726) 4.
29 *Relation* (1726) 4. This is the final sentence of the pamphlet.
30 Cysat (1969) 882–887; quotes on pp. 882, 884.
31 Di Somma (1641) 63.
32 Travagini (1673) 1.
33 Travagini (1673) 2–3. Bottari (1748) 98, mentions what Travagini had told him about his observations in Venice in 1667.

2

READING ON EARTHQUAKES

Explanations and interpretations up to the end of the Renaissance

The most common way how scholars in the early modern world knew about earthquakes, was via texts. Even when they had experienced one or two earthquakes themselves, their experiences would be heavily tainted by what they knew about these events from oral or written tradition. After all, earthquakes are rare events. To identify something that one sees happening as an earthquake, one has to know about these phenomena in general. Moreover, people's minds will normally be beset by panic, awe, and worry about friends and relatives, so that even direct observers seldom will be able to study the phenomenon accurately and dispassionately. Therefore, what people thought about earthquakes, to a very large extent consists of what they had read about them. A study of early modern ideas on earthquakes therefore should first of all investigate the existing traditions.

Earthquakes were well known in the classical world. Greece, Italy, and Asia Minor are all earthquake-prone countries. Classical authors, as inhabitants of those regions, could hardly ignore them. Thinking and writing about earthquakes was part of the classical tradition as it was transmitted to Europe. But new knowledge as well typically resulted from the study of texts. Increasingly in the sixteenth and seventeenth centuries, the body of ancient and medieval authorities was supplemented by a corpus of descriptions, histories, and contemplations on more recent events. Although written with an eye to the classical traditions, these descriptions often answered to more modern standards. Scholars increasingly took account not just of the ancient authorities, but also of these more recent descriptions. Early modern texts on earthquakes are not only the sources from which we learn people's ideas, but also the sources from which contemporaries got their inspiration for innovations in the theory.

Religious traditions

Earthquakes, even in countries where they are rather frequent, belong definitely to the category of the extraordinary and the uncanny. As such, nearly all cultures relegate them to the domain that we call religious. In myths, earthquakes are linked to the gods. So, Poseidon is called "earth-shaker" in Greek mythology. However, myths should not be read as textbooks of physics. In most cultures, the idea of physical causality is not strongly developed. Events are explained not with reference to natural principles, but rather to moral order. Mythological stories make sense of human life by referring to direct human experiences. By saying that earthquakes belonged to the domain of the religious, I do not mean that people looked for the cause of earthquakes in divine intervention rather than in natural processes. I first of all mean that they were allotted special significance in the overall understanding of the world.

In the early Christian world, earthquakes were frequently seen as special signs of God or manifestations of the divine order. However, authors describing earthquakes were seldom concerned with the causal explanation, or whether earthquakes were more "supernatural" than other phenomena in the world. The main point at issue was the moral significance. The idea of "natural" versus "supernatural" processes itself as it had been introduced by the philosophers of ancient Greece for a long time was far from universally accepted.

In Western Europe, the meaning of earthquakes was principally taken from religious writings, the Bible in the first place. There are several references to earthquakes in the Bible. They are mentioned in passing in Amos 1:1 and Zechariah 14:5. In 1 Kings 19:11–12, an earthquake is among the phenomena that precede the appearance of the Lord, and Isaiah 29:6 mentions an earthquake among the instruments with which God annihilates the enemies of his people. Several other texts in Isaiah speak about God making the earth tremble, such as 2:9, 5:25, 13:13, 24:18–19, and 64:1–3; likewise Job 9:6. In 2 Samuel 22:8 and Psalm 18:8, the shaking of the earth is mentioned as an expression of God's anger. Many other texts could be seen as relevant, for instance, Numbers 16:31–33, where it is recounted how the Israelites Korah, Dathan, and Abiram, who revolted against Moses, were swallowed by the earth.

However, the mentions of earthquakes that had most impact are those in the New Testament. Matthew 27:51 tells how an earthquake occurred during Christ's death on the Cross. The resurrection, three days later, was equally marked by a large earthquake and other wonders (Matthew 28:2). According to Acts 16:26, an earthquake shook the prison where Paul and Silas were confined, breaking their shackles. Last but not least, Luke 21:11 and Mark 13:8 mention earthquakes prominently among the signs that will announce the Last Judgement. The association of earthquakes with the end of times is further enhanced by several mentions in Revelation, notably Revelation 11:13. (This association with the Last Judgement exists in Islam as well, see in particular surah 99 of the Quran.[1])

In the Middle Ages or the early modern period, these biblical stories were a living reality. They served as models to understand and describe the world of everyday experience. When medieval chroniclers referred to earthquakes, they often borrowed their terminology from the Bible, especially Revelations.[2] The story of Korah and Abiram was the model for several similar stories. In the sixteenth century, Georg Agricola related it to several other cases from ancient history where humans were said to have been swallowed by the earth. He also reported about a case closer to home, where a girl had "recently" been swallowed by the subsiding earth, without an earthquake.[3]

At Christ's Passion, according to Matthew "the earth shook, the rocks split". People naturally looked for evidence of this in the landscape. The cleft in the rock of Golgotha was shown to pilgrims to the Holy Land from medieval times onwards.[4] The sixteenth century Catholic Church historian Cesare Baronio mentioned that, apart from Jerusalem, "in various other parts of the world as well, mountains have been split by that earthquake, as testified by solid traditions from the locals". He mentioned two rocks in particular, both in Italy: a cleft rock near the sea at Gaeta, and the Alvernine mountain in Tuscany. Both would be cited by other authors as well. A nineteenth-century author related that even in his time, the rock near Gaeta was the subject of special devotion by sea-farers.[5] Protestants had their own relics in the landscape. The Lutheran minister Sigmund Sack in a passion sermon in 1591 mentioned a common belief that the cleft rocks in the Harz mountains had been produced during Christ's Passion. The same is stated by a chronicle from 1613 and in some later works.[6]

Apart from the Bible, there existed a large body of other texts which used similar elements to make sense of the world: stories, not physical analyses. The classical (pagan) Greek and Roman historians had not been very interested in natural phenomena. They referred to earthquakes when they intersected with their narrative, but hardly as topics in their own right. Thucydides did briefly discuss the earthquakes and tidal waves of 426 BC, but his main reason for mentioning them was that they caused the Spartans to cancel their invasion of Attica.[7] A more flagrant example still is Livy's reference to a severe earthquake in Italy in 217 BC. Although this quake allegedly destroyed several cities, Livy just mentioned it for the fact that the soldiers who at that very moment were fighting the battle at Lake Trasimene did not even notice it.[8]

In late antiquity, however, historians became increasingly preoccupied with strange and ominous events. The fourth-century historian Ammianus Marcellinus included several descriptions of earthquakes and other natural disasters in his work. These had no direct impact on the political events he described and at first sight the descriptions appear as mere ornaments. However, they have a clear function in the narrative. As infractions upon the order of nature they function as omens, as comments on a supernatural level, that reveal the nature of human actions.[9] Late antique historians were not interested in nature for its own sake, but they did feel that natural occurrences, especially disastrous ones, conveyed a moral lesson.

The Christian authors of the Middle Ages followed along the same path. They were definitely interested in remarkable phenomena. Around 1200, Gervais of Canterbury explicitly demanded that apart from other things a chronist would record portents and miracles.[10] However, such events were told for their significance, not because of an interest in natural phenomena. The way authors referred to such events was highly dependent upon the particular genre. Saints' lives often mentioned earthquakes in imitation of the biblical ones, to underline the saint's imitation of Christ. Such references generally had no basis in fact. Chronicles, on the other hand, did strive for factual accuracy. The dates and times that were mentioned were normally accurate (locations less so) and the phenomena themselves were identified and discussed in the proper technical terms, that is, terms taken from natural history, not theology. Although strange phenomena were seen as portentous, there was no systematic attempt at interpretation. Only rarely authors showed a conscious didacticism.[11]

Still, events were included because they were considered meaningful and medieval chronicles abound with references to all kind of strange events. These were recorded not out of a curiosity for natural wonders, but as indications of the moral constitution of the time. Earthquakes and similar phenomena were elements in a general story about the world. They showed God's intentions with the world. By punishing sinners, humiliating the proud, proclaiming the future greatness of a newborn child, or vindicating holy men, they confirmed or re-dressed the moral order. The way to talk about these things was not in the form of a physical discourse, but as an *exemplum*.[12]

Chronicles of this nature lasted well into the early modern period. The chronicle of the monastery Hirsau, written by its abbot Trithemius, who is often regarded as one of the first Renaissance figures in Germany, abounds with the stories of earthquakes, blood rains, and other portents. So, under the year 1117, Trithemius relates how during a major earthquake, the water of the river Po in Italy rose so high that it stood as a cupola, suspended in the air, until it fell back with an enormous sound. At Cremona, a baby called to his mother and older brother ("against nature"):

> Stop preparing mortal things, you who will perish quickly! Behold, the holiest mother of God, the virgin Mary, stands before Christ's tribunal to avert with her supplications the judgement, that the judge intends for the world because of our sins![13]

A German broadsheet relates the destruction of Constantinople by an earthquake on 14 September 1509. Part of it simply reports on the damage in the city – the palace, the menagerie, the city walls, the mausoleum of the former emperor, and many towers all collapsed. It goes on rehearsing the destruction in other cities nearby, and tells of the high dignitaries that were killed. Part of it brings the familiar theme of the unheeded warning: it concludes with a story of an ancient Greek at the court of the sultan who had foretold the event.

However, the description of the destructions itself contains a message. Explaining what happened to the Church buildings, the pamphlet asserts: "Many of the churches of the Turks and the Greeks have collapsed, but nothing happened to the Christian churches." The Church of St. Sophia, which had been transformed into a mosque, remained unharmed as well; only a tower that had been added by the Turks after the conquest was torn apart from the main building and collapsed. Moreover,

> the chalk and mortar with which all the images of our lord Jesus Christ and his holy mother and other saints, to the disgrace of our faith, had been covered and hidden for a long time, now by such an earthquake, the chalk and mortar fell down from those images, as if removed artfully and purposefully, and they were seen as before.[14]

No interpretation of these events is offered, but their meaning is clear enough: God himself showed his displeasure with the Muslims; the Roman Church was the true Church, and protected by God. The news apparently spread widely. In 1510, The Bavarian historian Aventinus noted that he had heard a rumour that the city of Constantinople had been destroyed by an earthquake, apart from the Church of Saint Sophia.[15]

An entry from 1602 in the chronicle by J. Francus is also worthy of note:

> On 25 May, there was a terrible earthquake at the island of Cyprus. A big mountain rose up, out of which large flames were flying. When this abated, there was noticed a strange sea fish, which had a hand on its back. It was soon caught and sent to the Bassa in Natolia. Moreover, in the air was seen floating a blood-tainted cross, which came down and disappeared. The Turks got very afraid, fearing that some major harm may befall their empire.[16]

As we will see, early modern thinkers stood rather ambivalent vis-à-vis these interpretations. On the one hand, theologians would gradually distance themselves from a "magical" interpretation of nature and therewith from the exempla-tradition (although it remained strong for a long time). On the other hand, as the Bible would more and more take a central place in Christian thought, contemporary earthquakes were often regarded in the light of the Biblical examples.

Ancient philosophers

At the same time, there existed in the western world another intellectual tradition. The ancient Greek philosophers had started to explain the phenomena in the world from "nature", that is to say, from the properties of the things themselves, thereby deliberately rejecting the mythological explanations that described them in terms of acts of divine, anthropomorphic beings. In most cases, this was not intended as an expression of atheism, but it did imply a conception of the divine that was completely different from the ideas current among most cultures. The divine was an order that could be understood, not a power that could be manipulated by religious rituals. From the twelfth century onward, theologians decided that this philosophical conception of the world, if not of the divine, was well in agreement with Christian monotheism and they made Greek natural philosophy the foundation of higher learning.

The philosopher who gained most authority in the western world was Aristotle. Moreover, of all Greek philosophers whose works have been preserved, Aristotle happens to have the most elaborate discussion of earthquakes. The two other classical authors that deal at length with earthquakes are not Greeks, but Romans. The one is the philosopher Lucius Annius Seneca, the other the encyclopaedist Pliny the elder. Their basic theories about the cause of earthquakes largely agree with Aristotle's.

Aristotle dealt with earthquakes in his books on nature, specifically in his *Meteorologia*. His main question is the cause of earthquakes. First, he briefly discussed and refuted three earlier opinions on earthquakes of the philosophers Anaximenes, Anaxagoras, and Democritus. Thanks to his short discussion, the opinions of these earlier philosophers were known to early modern authors. Having refuted these, he comes with his own explanation. Under the influence of the heavens, the earth produces dry exhalations, in the same way that water when heated produces vapours. Whereas the watery vapours in the atmosphere are the cause of rain, snow, dew, etc., the dry exhalations are foremost felt as wind. (Aristotle rejected the theory that wind was just moving air, which had been defended by Democritus.) But not always do dry exhalations move so freely through the atmosphere. Sometimes, they are trapped within clouds; when thereupon they break out forcefully, they cause thunder and lightning. And sometimes they turn back into the earth. When they get trapped within subterranean cavities, they move violently until they find a way to break out. This violent moving is felt as an earthquake.

Aristotle explained how various phenomena follow quite logically from this explanation. A strong point is that circumstances that favour wind are conductive to earthquakes as well. Most earthquakes, especially large ones, occur in calm weather. Most major earthquakes occur at night, and those that occur during the day do so mostly around noon. They occur at places where the sea is full of currents and the earth is porous and hollow. They occur most often during spring and autumn and during rains and draughts. Of course, according to modern insights none of these observations holds water. To what extent Aristotle did refer here to common earthquake-lore as it existed among the Greeks, and to what extent he built on own (necessary limited) data-gathering is hard to assess.

In a next section, Aristotle sought further confirmation for his theory, partly from what was known from historical earthquakes (he must have collected accounts about these), partly from other phenomena. He was particularly attentive to signs and other phenomena that (as he claimed) accompany earthquakes. So, he claimed that before an earthquake occurs, the sun becomes nebulous and dimmer though there is no cloud, and when an earthquake occurs at dawn, there is often a calm and a hard frost. Another sign that according to him often heralds earthquakes is a fine long streak of clouds that appears in clear weather, either at day or a little after sunset. He also maintained that earthquakes often occur at an eclipse of the moon. Aristotle also briefly discussed some further attendant circumstances, such as subterranean noises, water bursting from the earth, and tidal waves. He explained that earthquakes are only local. How he explained these phenomena from his theory of exhalations needs not detain us here. What is relevant is that all these signs were common knowledge in the Middle Ages and early modern period.

Besides these naturalistic elements, however, Aristotle also presents some elements that are more animistic. He explicitly draws an analogy between the earth and the human body: "For we must suppose that the wind in the earth has effects similar to those of the wind in our bodies whose force when it is pent up inside can cause tremors and throbbings...."[17] Earthquakes were literally colics of the earth, an idea that would be repeated by later authors as well. For us, such an animistic explanation seems at odds with the seemingly more mechanical explanation in terms of exhalations. Medieval authors did not see much of a contradiction, but in later times scholars were more at pains to balance them.

The discussion of earthquakes in the scholastic tradition

The Aristotelian tradition obtained a dominant position in European learning from the twelfth century onwards. The newly founded universities accepted Aristotle's philosophy as a basic programme for learning. This included his books on meteorology. The work acquired textbook status and became the subject of a great many commentaries and paraphrases. One of the first, and most authoritative, commentaries was written by Albertus Magnus in the thirteenth century. Albertus not just slavishly repeated Aristotle's ideas but interpreted him rather freely, adding and complementing him on points of detail.[18] The meteorological tradition lasted well into the seventeenth century. These works were overall based on Aristotle, but some information derived from other authors could also be included. Especially later authors often tried to show off their erudition in this way.[19]

In this tradition, a standard way of discussing earthquakes developed, which was not only applied by scholastic authors, but came to be used in many sermons and histories as well. Likewise, the template was followed not just by Aristotelians but also by some authors inspired by more modern ideas. In medieval vein, the subject was discussed under various headings or questions. The authors might show their personal views in their answers, but they rarely diverted from the questions themselves.[20]

- The cause of earthquakes
  Most authors follow Aristotle. Earthquakes are treated among the phenomena of dry exhalations, like wind and lightning. Not infrequently, the

authors pass in review the four elements as possible causes: is water the cause (as Democritus thought)? is earth the cause (as Anaximenes thought)? etc. This offers the occasion to rehearse the views of these and other philosophers. There is often also a reference to the views of the Babylonians, who allegedly attributed earthquakes to the influence of the stars, in particular the three upper planets (Mars, Jupiter and Saturn). This information goes back to Pliny.[21] Other authors could be quoted from more obscure sources.

- A categorization of earthquakes

  This goes straight back to the ancient sources. Still, different sources use different categories. Most authors follow simply one of them. Those who follow Aristotle distinguish mostly between two kinds of earthquakes: *pulsus* and *tremor*. Notice that although based on Aristotle's work, in that Aristotle speak about the different ways earthquakes can be felt, he himself did not do this in the form of a classification. Pliny spoke of three kinds of earthquakes: *ruina*, *hiatus*, *pulsatio* (or *tremor*). Seneca too distinguished between three kinds: *succussio*, *inclinatio*, and *tremor*. Still another categorization could be taken from the pseudo-Aristotelian work *De mundo*, which distinguishes no less than eight varieties.

  Nearly all early modern authors follow one of these classifications. Many give several of them. This is still true for an eighteenth-century work as Chambers *Cyclopedia*.[22] Other authorities, however, do pop up now and then. Girolamo Cardano came up with a classification taken (without reference) from the historian Ammianus Marcellinus, who focussed on prodigious effects rather than physical aspects. The theologian Franciscus Torreblanca also preferred Ammianus, rather than Aristotle or any other philosopher.[23]

  Some authors discussed the classification in a more critical vein or came up with their own ideas. Georg Agricola explained the classifications according to Aristotle, Poseidonius, and Seneca and then tried to harmonize them. (Poseidonius mentioned two forms, which Agricola identified with the *succusio* and *inclinatio* of Seneca.) He felt one should make a distinction between simple and composite earthquakes. Aristotle's *tremor* and *pulsus* are both simple forms. The *pulsus* can be subdivided into *succussio* and *inclinatio*. There is one composite form of earthquakes, which he called *arietatio*. So all in all there are four types: *tremor*, *succussio*, *arietatio*, *inclinatio*. Either the *inclinatio* or the *arietatio* is the most dangerous.[24]

- Signs and ways to predict earthquakes

  Most of this is again based on Aristotle. Since the cause of earthquakes was felt to be in dry exhalations that also caused wind, it seemed obvious to look for a connection with the weather. Earthquakes happened when exhalations had turned inwards instead of causing wind on the surface. They were therefore felt to happen at quiet weather. Seneca affirms the same.[25]

  A sign frequently mentioned in the descriptions is a long thin cloud seen over the horizon. According to Aristotle: "In clear weather, either by day or a little after sunset, a fine long streak of cloud appears, like a long straight line carefully drawn, the reason being that the wind is dying down and running away."[26] This cloud not just turns up in general descriptions, but also in reports about specific earthquakes, with amazing persistence. Still in 1750, in an article in the *Gentleman's magazine*, George Berkeley wrote that he had heard from a witness of the earthquake of 1693 that "some hours before … he observed a line in the air (proceeding, as he judged, from exhalations poised and suspended in the atmosphere)".[27] Evidently, as the cloud is supposed to have preceded the earthquake, it is always with hindsight that people remember to have seen it. Another sign frequently mentioned is that before an earthquake, the sun becomes misty and dimmer.

A sign frequently mentioned is the muddling of water in wells. This sign is connected to a story from Pliny about the Greek philosopher Pherecydes (according to some reports the teacher of Pythagoras), who allegedly predicted an earthquake from this sign. The philosophers Thales and Anaximander too were sometimes reported as having predicted an earthquake. Some authors include other signs as well. Flaminio Mezzavacca talks among other things about lightning, bees and mice becoming restless.[28]

• When and where do earthquakes occur?
Again, because of the connection to the weather and the atmospheric conditions, it was felt that earthquakes happened preferably at specific times of the day. According to Aristotle, "[m]ost major earthquakes occur at night, and those that occur in daytime at midday, this being as a rule the calmest time of the day". Moreover, earthquakes occur most often in spring and autumn and during rains and draughts, "since these periods produce most wind".[29] (Pliny points out that these are the same seasons that lightning occurs most often.) These points are accepted by most authors on earthquakes, Aristotelians or not.

It was already noticed in antiquity that some places were more liable to earthquakes than others. Philosophers tried to come up with some systematization. Earthquakes were supposed to occur most in islands (but, according to Aristotle, are rare on island far out in the sea), near the sea-coast, and in mountainous areas. Moreover, they were supposed to happen at places where the earth was porous and hollow, and where the sea was full of currents. Caves and hollows were supposed to provide space for the exhalations, the sea was supposed to prevent the exhalations from escaping. Aristotle just gives a general description, but modern authors normally come up with concrete examples, like Sicily. Not all examples are completely reliable. Pierre d'Ailly mentions as examples of spongy places near the sea, which are prone to earthquakes, not just Sicily but also Egypt, which is not particularly earthquake-prone.[30] Garcaeus mentions in particular a coastal region which, as he claims, because of its many caves is called Holland – again, a country where earthquakes, and caves, are in reality very rare.[31] Interestingly, the (mistaken) etymology to which Garcaeus appeals seems to work only in Dutch (hol = cave). The confusion may also partly be based on an association of earthquakes with inundations, which indeed were frequent in Holland. This seems to be the case with Luco Maggio, who repeats Garcaeus when talking on inundations following earthquakes.[32]

- Remedies against earthquakes
Even today, it is pretty hard to come up with remedies against earthquakes. Seneca explicitly denied that there were any: earthquakes cannot be predicted; there is no place where you can hide from them (certainly not in the earth); and you cannot escape from them, because wherever you go, you may be struck by an earthquake. Still, it is understandable that people did not simply acquiesce in this. Nearly every treatise has a section on remedies against earthquakes.

   Pliny came up with some practical advice about what parts of buildings were most safe in the case of an earthquake: "The safest parts of buildings are arches, also angles of walls, and posts, which swing back into position with each alternate brush."[33] Also, he was read as saying that vaulted buildings were sturdier. Pliny also noticed that regions with many wells and other conduits were less shaken, since they supplied an outlet for the exhalations caught underfoot. Hence, many earthquake texts recommend the digging of deep wells as a prevention against earthquakes.

- Effects of earthquakes
Here, one can of course make a long list: cities devastated; clefts in the earth; disappearance of islands; rising or flattening of mountains; water flowing from the earth; fire emerging from the earth; all kinds of sounds. An effect not mentioned by Aristotle but found in about all medieval and early modern textbooks is plague. Air, which has long been enclosed in the bowels of the earth and has become rotten and poisonous, escapes to the surface and causes disease. The point is already made by Albertus Magnus, who alludes to an experience he witnessed: when a well was opened after a long closure, the air killed several people.

- The final point is generally a list of earlier noteworthy earthquakes. This list of course was frequently updated. It also gave ambitious authors occasion to show off their erudition by including earthquakes mentioned only in very obscure sources. Still, the examples mentioned by the ancient sources were seldom lacking.

Wonders in Aristotelian philosophy

The professors who introduced Aristotle into the medieval universities were theologians and it was mainly for theological reasons that Aristotle's work was accepted. Aristotle looked for the causes of everything in the world only in nature itself. That is, nature was a self-contained entity; it was not guided, directed or caused by something outside or above. It might not appear immediately apparent why this view of the world would appeal to theologians. Still, the medieval scholastics took this aspect of Aristotle's philosophy quite seriously. Evidently, they believed that the world had been created by God. But then, they maintained that God had left his creation to its own inherent powers. Although God kept the freedom to intervene directly into the world, he did so only on rare occasions.

Concretely, this meant that they believed that unusual phenomena like comets, blood rains, eclipses, earthquakes, and other strange and frightening phenomena, were no manifestation of the supernatural, either in the form of divine miracles or otherwise, but could be explained naturally. In this respect, Aristotle's explanations were of great use. Medieval scholastics even included new phenomena into their discussions of meteorology, which they then explained in a like way. Albertus Magnus already included the "flying dragon" or "firedragon". Like comets, this too was a form of dry exhalation that caught fire in the atmosphere. In this way, according to Albertus, one could explain how simple people believed that they had seen dragons.[34] In the same vein, scholars explained prodigious rains, such as rains of blood, fish, or iron, from natural causes. In the later Middle Ages, will-o'-the-wisps were included as meteorological phenomena as well.

It is noteworthy that such explanations remained not limited to philosophical textbooks. They also entered medieval chronicles and histories. The so-called chronicle of Saint Pantaleon under the year 1235 gives a detailed description of a celestial phenomenon and comments: "Although these circles by the people were considered as a prodigy, their causes have been explained by Aristotle and Seneca in the meteorology."[35] The leading theologian Thomas Aquinas came to speak of earthquakes in his commentary on Psalm 18. He explained the cause of earthquakes following the Aristotelian explanation. In his view, it was said only metaphorically that earthquakes were an effect of divine wrath; their intention to move people to penance should be understood in a mystical sense. Of course, the first cause of earthquakes, as of any other phenomenon, was still the divine will.[36]

Aristotelian philosophers explicitly aimed at debunking "superstitious" beliefs. This appears as part of a general re-interpretation of the division between the sacred and the profane. The divine was not to be met in specific objects or events, but in the doctrine of the Church. Theologians consciously opposed the world view of the early Middle Ages, when people had felt a much greater presence of the Divine in their daily lives. God was now placed at a greater distance from humans; he could not be approached by mere human ritual. The sacredness of relics, of images, of Church buildings was de-emphasized. They did not have inherent powers that worked automatically. God was transcendent and could make his presence felt everywhere. Theological considerations were guiding here, especially regarding God's transcendence and omnipotence, but there may have been more worldly ones as well. When miracles were exceptional, the only way to have access to the Divine was by the mediation of the Church. The Church in the end decided what counted as a miracle and what not.

From this point of view, extraordinary phenomena could not be given any inherent meaning. At the same time, these philosophers were unwilling to contradict the classical authorities who interpreted them as harbingers of evil. Typical is the way they dealt with comets. Traditionally, comets were seen as heralds of all kind of evil – plague, famine, war, and the death of kings. The philosophers did not doubt the reports themselves, but felt the evil effects had natural causes. Comets, according to Aristotle, were caused by dry exhalations, which caught fire in the upper layers of the atmosphere. It was easy to conclude that these same exhalations also poisoned the air, causing plague. Plague thus followed comets in a natural way. Because Kings had a more tender constitution than common people, they were more liable to die. An alternative explanation claimed that the exhalations were raised by the influence of the planets, in particular the upper planets, Mars, Jupiter, and Saturn. The same influence also strengthened the bile in humans, giving rise to anger, conflicts, war and thus indirectly again to the death of kings.[37] In this way, philosophers managed to harmonize their naturalism with traditional beliefs.

The effects of earthquakes were rationalized in a similar way. The fourteenth-century author Konrad of Megenberg wrote in his encyclopaedic *Buch der Natur* [Book of nature] that many wondrous effects were wrought by earthquakes. Notably, from the vapours rising from earthquakes, people and animals were turned into stone, in particular salt stone. This happened most often in mountainous areas and around salt mines. Konrad referred to the authority of Avicenna and Albertus Magnus, but also to a story he had heard from the chancellor of Austria:

> that on a high Alp in Carinthia fifty people and cattle had been turned into stone, and that the milkmaids were still sitting under the cows with a glove, just as they had sat when they turned both into stone.[38]

A miracle story, modelled after the biblical story of the wife of Lot (Genesis 19:26), is here rationalized by means of the theory of exhalations.

In later writings, this event is identified with the earthquake of Villach, in Lower Austria, in 1348. The Bavarian historian Aventinus wrote in his chronicle:

> Conrad of Meidenberg, a prominent philosopher and mathematician of that time, relates that over fifty people, villagers who were milking their cows, died and stiffened with their animals in Carinthia. Their bodies were turned by the terrestrial exhalation into salt statues and these statues have been seen by him and the chancellor of Austria.[39]

Notice that the stone in the original story is now explicitly salt, in case someone might miss the analogy to the biblical precedent, but the reference to exhalations is maintained. The event is reported by many other chroniclers, both Catholics and Protestants, well into the seventeenth century.[40]

The relation between the physical explanations of the philosophers and the theological interpretation of the philosophers was not always clear. Medieval philosophers too were generally clerics. They might reject popular superstitions, but they did not want to oppose the principle that God can and does act in this world in miraculous ways. So, the plague was explained in a natural way, but in terms that were strongly reminiscent of the religious world. Natural remedies were not efficacious in their own right: they were gifts of God and only God could grant (or withhold) their efficacy. Nature was not autonomous. The language of natural explanations remained part of the overall symbolic and religious interpretation of the world.[41]

Related phenomena: volcanoes and landslides

A question that needs to be addressed here is the relation of earthquakes to similar phenomena, notably volcanic eruptions and landslides. Up to the nineteenth century, people had no idea of seismology and the distinction between such phenomena remained necessarily blurred.

As for volcanoes, it is striking that they are not discussed by ancient philosophers. Whereas most people in the Mediterranean world would have first-hand experience of earthquakes, volcanoes are confined to specific localities. Eruptions were regarded as a specific property of mount Etna, not a general property of the earth.[42] In other cases, volcanic eruptions seem to have been considered as a kind of earthquakes. So, Aristotle described an "earthquake" in the Aeolian islands, which based on his description we would definitely classify as volcanic:

> in this island part of the earth swelled up and rose with a noise in a crest-shaped lump; this finally exploded and a large quantity of wind broke out, blowing up cinders and ash which smothered the neighbouring city of Lipara, and even reached as far as some of the cities of Italy.[43]

Aristotle is clearly considering the phenomenon through the lens of his earthquake-theory. As earth is a cold and dry body, and its natural place is the centre of the world, it was evidently hard for him to accept that terrestrial phenomena like earthquakes could be caused by fire. Indeed, the fire there undeniably is, he considered as an effect of the exhalations.

Some theory of "burning mountains" was offered in the Middle Ages by a pseudo-Aristotelian work, *De proprietatibus elementorum*.[44] The work appears to be Arabic in origin, but does use ancient sources now lost. It explains eruptions of fire from the earth not as a form of earthquakes, but as occasioned by large quantities of subterranean sulphur catching fire. Merely from this description, it was hard to get a good idea about volcanoes, especially if one had never seen one. Albertus in his commentary to this text referred to a burning mountain near Liège (present day Belgium) from which by rain came smoke "as from a furnace".[45] Georgius Agricola when discussing "burning mountains" in his work, considered a *mons carbonum*, a charcoal mountain, in Meissen as on a par with Etna or Vesuvius. Still in 1666, the scientific amateur Alexander Achilles claimed that volcanoes and hot springs were just coal layers that had been lit by lightning.[46]

After people had obtained more knowledge about the world, they came to realize that volcanic eruptions were general phenomena and that Aristotle's meteorology in this respect was wanting. By the early modern period, eruptions of Vesuvius, Etna, or any other volcano were normally treated as a class of events in their own right. Still, since neither earthquakes nor volcanic eruptions were well understood and most information came from hearsay, there was still much overlap possible. When Aristotle described something as an earthquake, later authors were not going to correct him. Another example of this confusion is the eruption of the Monte Nuovo, the "New Mountain", in the Phlegraean fields near Naples. This mountain emerged as the result of an eruption in 1538, growing to a height of about 100 meters practically overnight. The event drew wide attention and almost all reports describe it as an earthquake.[47] This interpretation seemed obvious because classical authors explicitly mentioned the flattening or emergence of mountains among the effects of earthquakes.

As for major landslides, distinguishing them from phenomena that we would classify as seismic was still harder. We are not talking here of simple avalanches, which were of course common enough, but of what in German is called *Bergsturz* – fall or collapse of a mountain. When the volume of falling rock is more than a million cubic meters (a cubic hectometre), the mass behaves like a viscous fluid, coming down with high velocity (up to 400 km/h) and flowing out far and wide. Such landslides may wipe out complete villages, killing people and cattle in an instant.[48] One such case happened in 1584, when the village of Yvorne (canton Vaud, Switzerland) was buried under a collapsing mountain. The collapse actually appears to have been triggered by a series of real earthquakes. A similar disaster happened in 1618 in what is now northern Italy, but was at the time subject to the Grisons. An area of about 1.4 square kilometres was buried under 20–30 meters of rubble; the village of Piuro (Plurs in German) and the smaller village of Chilano (Schilan) were completely destroyed.[49] People did realize that these events in their strictly local effects were distinct from normal earthquakes and even that one could point at specific causes – in the case of Piuro, the mountain had become unstable by human mining activity. The Italian author Hieronymus Cardano already in 1550, in his book *De subtilitate,* had a separate discussion of small "earthquakes" caused by the erosion of mountains.[50] Still, throughout our period such events are invariably termed earthquakes and treated as such. Only in the eighteenth century, the distinction would become common knowledge, propagated by authors like Johann Jakob Scheuchzer.

As a consequence, the events of Yvorne and Piuro will be discussed in this book on a par with real earthquakes, as this corresponds to the understanding of the contemporaries. The same is true for the eruption of the Monte Nuovo. However, eruptions of Etna or Vesuvius will be referred to only incidentally, when they intersect with the history of ideas on earthquakes proper. This is partly a way to keep the material manageable, but can also be defended on the basis of contemporary understanding.

Notes

1 For early modern Islamic attitudes to earthquakes, see Akasoy (2007); Schulze (2004); Tahir (1974).
2 Draelandts (1995) 60–61, 71.
3 Agricola, (1546, *De natura*) 153. This is supposed to have happened "Vratislavij in Lygijs, quos Slesitas nominamus", which I take to be Vratislavice nad Nisou (or Maffersdorf) in Bohemia, near the Saxon border.
4 Brefeld (1994); see pp. 200–201, nr. 131, and 103–104 for some quotations from the manuscripts. In fact, the item is not in the original Jerusalem "tourist manual" as reconstructed by Brefeld. She argues that it is among the items which were part of a guided tour through the city (ibidem 112). Still, the item occurs in quite a number of manuscripts. It is still there in sixteenth-century descriptions of the Holy Land.
5 Baronio (1738–1746) I, 155. Meurer (1587) 239. *Fernerer Bericht* (1627) A3v (on the rock at Gaeta). Magnati (1688) 4–5: "come nella morte del nostro Redentore vi fù un tanto gran Terremoto, che spezzò sino alle pietre, facendo testimonianza il Monte Alvernia nella Toscana, e la Città di Gaeta nel sudetto vostro Regno di Napoli, il Monte Calvario in Gerusalemme, conforme lo testifica Eusebio, Baronio, e San Cirillo Vescovo della sudetta Città di Gerusalemme," von Hoff (1840) 170.
6 Sack (1591) 121v: "…und wird dafür gehalten / das auch die Klüffte am Brocken Berge zu der Zeit ohne zweifel geschehen sein". Binhard (1613) 4. *Unglücks-Chronika* (1692) (no page numbers) refers to Binhard. Disputation Bremen (1684 May 3) refers to Sack. Höpffner (1691) 14.
7 Thucydides, Peleponnesian war, book III, chapter 89.
8 Livy, *Ab urbe condita*, book XXII, chapter 5, at the end.
9 den Hengst (2010) 240, 246–247. Cf. Graf (2010) 108: "ancient historians … liked to explain catastrophes in theological and moralistic terms. Theirs was not science, but a discourse about the forces in human history…" Weber (2015) 60–65.
10 Lehner (2015) 10–11.
11 Draelandts (1995), on hagiographic texts: 101; on chronicles: 79, 142–143, 146–147. See also Lehner (2015), passim. For the Byzantine world, cf. Ducellier (1996) 66, 72–74.
12 On the concept of *exemplum* in this respect, see Berlioz (1998) 33–55.
13 Trithemius (1690) I, 364. For other mentions of earthquakes and related prodigies e.g. I, 461, 502, 532, II, 230, 233, 649. The work was first published in 1559.
14 *Erneuerung* (1509). I tried to keep the original syntax.
15 "Fama fuit in principio huius anni Bizantium terremotu corruisse praeter templum Sophiae superiori anno in Septembri factum." Aventinus (1884–1908) VI, 13.
16 Francus (1602) 74.
17 Aristotle (1978) 209 (Meteor. II 366b).
18 Oeser (1992) 17–18.
19 On the meteorological tradition, see Martin (2011); Céard (2013); Ducos (1998); Heniger (1960); Hellmann (1917). For earthquakes, see also Adams (1938) 399–415.
20 E.g. Jacobus Amsfordensis (1513); Pierre d'Ailly (1506); *Commentarii Collegii Conimbricensis* (1600); Vicomercatus (1565); Nipho (1560); Wildenbergius (1585); Mezzavacca (1692) (on the earthquake of 1672).
21 Nat. Hist. II, 191.
22 Chambers (1741) s.v. 'Earthquake'.
23 Cardano (1663/1966) III, 16. Torreblanca (1623) 224–225. The reference is to Ammianus Marcellinus book XVII, chapter 7, par 13–14. See Ammianus (1968) 228–229.
24 Agricola (1546, *De ortu*) 31; (1546, *De natura*) 150–151. In *De ortu*, in the text the *succussio* is called *concussio*, but this is corrected in the "castigationes" at the end of the book.
25 Aristotle Meteor. II 366a. Seneca II 12, 2.
26 Aristotle (1978) 213–215 (Meteor. II 367b).
27 Berkeley (1951) 255.

28 Pliny, Nat.Hist. II, 191. Mezzavacca (1692) 263.

29 Aristotle (1978) 207, 209 (Meteor II, 366a, b).

30 d'Ailly (1506) 13v.

31 Garcaeus (1584) 389v: "Sic Batavorum tractus, aliaque loca cavernosa, recte à cavernarum multitudine Hollandia dicta, huic malo saepe sunt obnoxia." The statement was repeated in disputation Wittenberg (1607) 21–22.

32 Maggio (1575) fol 54v.: "Holland, so named in that language [*cosi detta in quella lingua*], according to the cosmographers, because in that country the earth is very cavernous, and can generate and contain many exhalations…".

33 Pliny (1979) 329. (Nat. Hist. II, 197).

34 Albertus Magnus (1651) 24: "Trabs autem ista cum expellitur si obviat ei frigus, incurvatur, & apparet ad modum draconis flexuosi: & hoc est quidem quod quidam dicunt se vidisse dracones volantes per aërem, qui ignem vomerent."

35 Lehner (2015) 154.

36 Udías (2009) 42. On the relation between naturalistic and supernatural explanations of earthquakes in the Byzantine world, see Dagron (1981); Ducellier (1996) 62–63.

37 Cf. Céard (2013) 39–40; Schechner (1997) 91–103; Ducos (1998) 403–404.

38 Konrad of Megenberg (1861) 112. See Weber (2015) 43–45.

39 Aventinus (1580) 633. Also in Aventinus (1884–1908) III, 462.

40 E.g. it is repeated almost verbatim in Eckstorm (1620), in the list of earthquakes that he appended to his history of the destruction of Piuro. Also in Babst (1599); Nuber (1655) 36; Höpffner (1691) sect. 67.

41 Smoller (2000); Esser (1997).

42 Cf. K. Taylor (2016) 118.

43 Aristotle, Meteor. II, 367a; Aristotle (1978) 211.

44 The text, with commentary by Albertus Magnus, has been printed in Albertus Magnus (1980) 49–104.

45 Albertus Magnus (1980) 99.

46 Agricola (1546), *De natura eorum*, cap. 17; *De ortu*, 34, 35. Achilles (1666) section 11.

47 On this event see Scarth (1999) 42–55.

48 Hauer (2009) 14–15.

49 An earlier case is the collapse of Mount Granier in 1248, discussed by Berlioz (1998).

50 Cardano (1611) 152.

3

WRITING ON EARTHQUAKES

The available information

What people knew about earthquakes was in the first place what they read in ancient authors. Still, the fact that earthquakes were happening also in their own time could not possibly be ignored. Whatever the authority of ancient philosophers, Church historians, or the Bible, in the end, the notion of earthquakes went back to real experiences and observations; experiences that never failed to make a deep impression. Information on these was therefore eagerly sought.

If in course of time, the ideas on earthquakes changed, not the least of the reasons was a continuous flow of new information. In most places of the world, earthquakes are rare events. At the end of the Middle Ages, apart from what they had read in ancient texts, many people's knowledge of the phenomenon may have been based on some vague stories about one or two earthquakes. Since that time, the invention of the printing press, the rise of literacy, the extension of networks of trade and diplomacy, and last but not least, the rise of postal services greatly improved people's access to knowledge about the world.[1] Around 1700, a host of authors had been collecting and compiling information and much of this now was available in printed form. True, not all information was reliable. Still, its sheer mass confronted both scholars and the literate public in general with very different problems than earlier generations.

It is therefore essential for our purpose to get some idea of the information on more recent events available at the time. Initially, most information of course was transmitted orally. We may safely assume that whenever earthquakes happened, they were the talk of the town. Bartholomeus Keckermann travelled through Switzerland in September and October 1601, just after the great earthquake of Unterwalden had happened. "I found that not only the mountains and valleys had been shaken, but also the human minds. Wherever I stayed was frequent talk of the earthquake, and a recounting full of religion and full of pious motion and fear."[2]

In a mainly oral culture, the memory would be preserved by stories, songs, and rituals. Scheuchzer noted in 1706 that the tradition of distributing bread and clothings to the poor on the anniversary of the 1356 Basel earthquake was maintained even in his day.[3] Rasch in 1582 gave, allegedly from oral tradition, a poem on the Villach earthquake of 1348.[4] But many people would put their experiences to paper or search for a more material souvenir. The widely felt earthquake of November 1692 in the Low Countries was mentioned by John Evelyn in his diary, even though he had not sensed the motion himself.[5] In the Netherlands, several people noted the event on the flyleaves of their family Bibles; another made a note on the flyleaf of a copy of a Dutch translation of Descartes' "Principles of Philosophy".[6] Whereas the former seem foremost concerned with details that can be religiously interpreted, the latter just gives the details about the time, the weather and the effects on people's health. After many earthquakes, commemorative poems were printed and sold.[7] In other cases, medals were struck.[8] More lasting were the inscriptions in stone that were afterwards put up in afflicted cities.[9] Such souvenirs were certainly instrumental in keeping earthquakes a living reality even at times that they were not felt, but for the rest were hardly informative. In most cases, they spoke more of human emotions than of what actually happened. Moreover, they were only concerned with local events.

For getting information on what happened farther away, letters were the obvious medium. People would want to tell friends and relatives what had befallen them or, on the other hand, they would want to inquire after friends. So, the botanist Carolus Clusius wrote in November 1590 from Frankfurt to his friend, the librarian Hugo Blotius in Vienna:

> There has been a large earthquake in your parts last September. No harm happened to you? I am very concerned about our Lewenclavius, as I have heard that the house where he was living has collapsed. Please write me everything....[10]

Such notes are of a principally private nature, but they were soon to be supplemented by more public reports. Not all of these were intended to convey information. Sermons would first of all draw a moral lesson, poems served as a souvenir, and so on. But as there was a real thirst for information, the immediate aftermath of a spectacular event regularly saw the appearance of news-sheets and pamphlets. Indeed, pamphlets on important events are among the first products of the printing presses.

Newsletters and pamphlets

There is no clear boundary between immediate reactions with a private character, and these printed pamphlets.[11] Many pamphlets appear in the form of letters sent by some person who witnessed the disaster. One German pamphlet on the earthquake of Ferrara in 1570 is titled: "An extract and translation of a letter of

21 November of this 70th year...." The text seems to indicate that the author was close to the Ferrarese court. A pamphlet from 1542 gives a report that concludes by stating: "This I had to tell you in a hurry. The mail goes off."[12] A pamphlet on the disaster of Piuro in the Alps refers explicitly to letters as trustworthy: the news is not just confirmed by general street rumour [*ein gemein Gassengeschrey*], but also by letters from prominent persons. The author confirmed that he had seen several of such letters, in German, Italian, and Latin, quoting four of them in full.[13] The earthquake of Apulia in 1627 was the subject of a pamphlet first printed at Rome. The Bolognese author explained in the introduction that initially, he had hesitated to publish the work, because

> the reports that came in disagreed so much among each other, that I did not know where to start or whom I should believe. So, I was ready to give up the project. But since several good friends have shown me credible letters, whose reliability cannot be doubted, they induced me to have this horrible story printed.[14]

In some cases, referring to letters was probably just a rhetorical ploy to give credence to the report, but in most cases, they must have been genuine. Printers had to get their material somewhere and letters from eyewitnesses or people on the spot were an obvious source. There were no professional press agencies, so, if a publisher did not want to make something up, he either had to copy some existing pamphlet or news-sheet, or have recourse to the more private channels that were used to convey news. Many letter-writers, especially scholars, probably were aware that the addressees might circulate their letters among a wider group of people, especially when they contained important news.[15]

The Swiss cleric Johann Jakob Wick in the years 1560–1571 compiled a huge collection of reports of events happening in his time. His collection includes a large number of printed pamphlets, either in the original or in handwritten copies, but also many reports that were copied from letters, as well as news that he learned by word of mouth, either from his fellow citizens or from travellers. Many of the letters had originally been addressed to his learned friends at Zürich, who knew about his project and were happy to give him access to their correspondence. Others were by more obscure persons. So, he included a copy of a letter on the earthquake of Ferrara of 1570 by a certain Jakob Held, secretary of the ducal guard in that city (many Swiss served as hirelings abroad). He noted that this letter had originally been sent to Barbara Kramerin, the wife of Thomas Eberhart the elder, trumpet player.[16]

Wariness about the accuracy of the reports was not always unfounded. A pamphlet on the earthquake of 1580, published at Paris, is based on a letter from Calais, but the letter writer is rather short on what happened in that city. Rather, he tells what reportedly has happened at sea and what he has heard from travellers arriving from England, which gives occasion to some rather tall stories.[17] One can see why people preferred letters from reliable persons over common rumours.

That stories were really made up was rare, but did happen. A flagrant example is a pamphlet on an earthquake and inundation at Reffel in Tanzgaw in November 1590, which according to the colophon was published in that place. The pamphlet relates all kinds of other portents that happened there as well; the earthquake, although allegedly pretty catastrophic (more than three hundred and eighty deaths), takes only a small place. The author refers to prognostications, prophecies and to Ptolemy. A problem with this pamphlet is that neither Reffel nor Tanzgaw are known place-names. The whole looks like a hoax rather than a real report. Apparently, the publisher wanted to profit from the interest in portents in the wake of the earthquake of Vienna and unable to lay his hands on any real material just made something up.[18]

Equally spurious is a pamphlet published in London in 1673 by someone with the impressive name of Leopold Wettersteint of Hodenstein. It is allegedly translated from the Dutch and claims to be the report of a major earthquake in Germany, Hungary, and Turkey, with many clefts in the earth that swallowed complete villages and erupted fire. The report is mostly based on the information of an (unnamed) imperial envoy who had just returned from the region. This quake is not recorded elsewhere, Mr Wettersteint is not otherwise known, and a Dutch or German original of the text cannot be found. Again, this is clearly a hoax.[19]

It is important to realize that pamphlets were only one element in this network of information exchange. Much information was shared orally or by letters, and without these largely invisible channels the printing press would have been pretty helpless. In most cases, these manuscript copies have disappeared, but there are still a few examples. The destruction of Yvorne in Switzerland in 1584 is described in a contemporary Latin manuscript of two pages, which claims to give the text of a letter from Bern dated 27 March 1584.[20] The text agrees in large part with a German pamphlet on the same event, published at Strasbourg in the same year.[21] This appears to be an expanded version of the same letter from Bern, although the pamphlet does not refer to any original letter. Compared to the Latin text, the German pamphlet has an extensive addition, which is not on the disaster itself, but gives a theological interpretation. This replaces a few lines in the Latin text on the repercussions of the quake at other places. The pamphlet is about double as long as the manuscript. Only one of its four pages is devoted to a description of the actual event.

In being turned into print, the reports changed character. Letter writers typically gave short and factual information. A printed work normally had more pretentions. Most often, it put the text in an interpretative framework and exhorted to piety. This may have to do with commercial considerations of the publisher, but also with the fact that, as will be discussed more extensively later on, in this era of Reformation and Counter-Reformation, the print medium was used as a medium of propaganda. One should not forget that products of the printing press were subject to censorship by the ecclesiastical and secular authorities.[22]

Printers used manuscripts, but in other cases, printed reports made it back into manuscript form. A manuscript chronicle by Johannes Wassenberch, a cleric in Duisburg on the lower Rhine, reports on the earthquake at Constantinople in 1509, repeating almost verbatim certain phrases that can be traced to the published broadsheet on the event. It is possible of course that both go back to a common manuscript source, but it seems more likely that Wassenberch directly or indirectly had access to the broadsheet.[23] Of course, such news might also be further propagated by word of mouth. As stated above, the Bavarian historian Aventinus reported on a rumour that he had heard on this same earthquake.

There are various pamphlets in German on a destructive earthquake in Savoy in 1564. The oldest one may be a single leaf, originally with an illustration, printed at Nuremberg in 1564. It has the form of a letter by a certain Francesco Mogiol, written at Nice 17 August 1564, and directed to his lord Jacomo Solomon. A short epilogue, which is not part of the letter, calls to penitence and puts hope in God's grace.[24] The same letter was published at Augsburg in the following year, 1565. It repeats the earlier text, including the epilogue, word for word.[25] Other pamphlets abandon the letter format and omit the name of the author. One such version of the text was published at Prague with an approbation of the bishop. It even omits the date.[26] Another one, published at Dillingen, gives the report itself on four pages, but has it accompanied by an introduction of five pages, wherein the author warns for God's anger. The epilogue, however, is lacking. This pamphlet still claims that it is a report from Nice, dated 18 August 1564.[27] Moreover, although the latter two pamphlets closely follow the same narrative, the wordings are different from the Nuremberg and Augsburg pamphlets, as well as from each other. It would seem that one and the same text circulated, initially probably in Italian, and was translated and (re)printed at various places, whereby each editor made his own choices. The new formulations are sometimes slightly more sensational.

It is not always clear whether a text derives from a manuscript or from a now lost pamphlet. With the development of the book trade, pamphlets and other works became more widely available. Instead of relying on manuscript sources, in the course of the seventeenth century, it became increasingly common for printers to rely on other pamphlets. Pamphlets with major news items were reprinted at other places and translated into other languages. In such cases, publishers would often emphasize their faithfulness to the original, putting on the title page things like: "following the original as printed in Vienna".[28] When the Paris printer Jean Coquerel printed a pamphlet on the earthquake of 1580, this was immediately reprinted at Lyon, Troyes, and an unknown third place.[29] A 1603 report on an earthquake in Constantinople was "originally printed in Vienna in Austria by Hans Schneideler, and reprinted at Magdeburg by Johan Bötcher", as the latter announced on his title-page.[30]

The news of the catastrophic earthquake of Calabria of 1638 was brought to the world in an Italian pamphlet from Rome, which then was translated and published in various other places. A Spanish translation was published at Barcelona and a German one at some unknown place. Both are small works of eight pages. The report is mainly factual and lists town by town the number of victims and other notable events. Here too, there is some variation in the introductory and final paragraphs. Another German translation, published at Breslau, presents itself as a "received report and continuation" and is expanded to twenty pages. It includes still another description of the earthquake, as received from Leipzig, as well as news of a new eruption of Vesuvius and some notable prophecies.[31]

Apart from their function of directly informing the public, pamphlets also were a source for the authors of larger works of history or natural philosophy. So, the earthquake of Ragusa was initially announced to the word in a pamphlet in Italian, printed at Venice in 1667, that was then translated into German and English. In 1677 a version of this text was included, together with some other sources, in volume ten of Matthäus Merian's *Theatrum Europaeum*.[32] In this way, pamphlets served not just the curiosity about current affairs, to be discarded after having fulfilled their function. The events they described became part of a growing body of knowledge about the world.

Histories

Short news sheets were not the only form wherein news about earthquakes was conveyed. Especially in Italy, where earthquakes can have really devastating effects, after a major quake, some authors might decide to write a full history, based on a more or less thorough investigation. In many cases, they had to fall back on the same method as the authors of pamphlets. Authors had to get their information from eyewitnesses and people who were familiar with the local situation. That again implied mostly writing and receiving letters.

Alessandro Burgos, a Franciscan who at the time was a professor of philosophy at Messina (later he was to become bishop of Catania), wrote a history (in the form of a letter) of the earthquake of Sicily of 1693. His book is something of a hybrid between a pamphlet and a real history. It is short, some twenty pages, and must have been written in some haste as it was published still the same year. For his information, he clearly relied on letters and other accounts. (Messina itself had not suffered much.) In the introduction, he rhetorically asked how to discern the truth amidst this misery and chaos. "But after we got certain informations on such a lamentable tragedy from letters that came in from the devastated places, we wanted to sketch in a few pages (...) the following report."[33]

Burgos relates the destruction of the convents at Paterno "according to a letter by Sir Alessandro Moncoda". For what happened in Caltabiana, he refers to a letter by the marquis of Francofonte, who had narrowly ("miraculously", according to Burgos) escaped with his life. About Chiaramonte, he tells that "by letters one has understood that at the collapse of the convent of the Minorites (Conventual Franciscans), two clerics have been buried. The number of dead is still uncertain". Noto, according to a messenger who arrived from there, had been completely ruined. Burgos gives a list of destroyed buildings. He notes that many people, among them a number of noblemen, have remained dead, and concludes: "On the clerics, one has no accurate information yet". Auolo has been destroyed "according to letters that have arrived from those parts".[34]

Like Burgos, many authors were clerics. They got their information to a large extent from fellow clerics. Communication among the clergy seems to have been surprisingly efficient, even during major crises. The clerical outlook shows in the facts that are described. The focus is largely on what happened to churches and religious buildings and to clerical persons. Besides, some mention is made of important local notables. The rest of the population mostly are reduced to anonymous numbers, apart from some spectacular cases. So, Francesco Grimaldi, on the authority of "a most trustworthy priest from our order", tells a story of a peasant who trapped between ruins remained suspended in the air for 27 hours.[35]

Local historians sometimes felt the event important enough to collect information and write a short history. Such histories could be used by others for information. Johann Huldrich Ragor explained in 1578 that an earthquake in Switzerland in 1534 (during which he was born) had been diligently described by his father, Heinrich Ragor, at the time a minister at Windisch. An earthquake at Bern in 1569 had been described by Master Benedict Martin. These works are otherwise unknown. In all likelihood, it concerns manuscript histories that were never printed. Johan Baptist van Helmont in 1640 referred to a manuscript history by the curate of the Church of St. Mary at Malines. It is hard to establish how prevalent such local histories were, or how detailed.[36]

In some cases, we have more information. Renward Cysat was preparing a major chronicle of Lucerne and the Swiss Confederacy when the earthquake of 1601 occurred. The book never appeared, but his notes have been preserved. He naturally wanted to include a description of the earthquake and he diligently sought out information. His notes are not just based on his own experience, but also on the facts "as I by diligent inquiry from my friends and most of the citizens learned to be true (as I myself was not at home at the time)".[37]

Writers of histories took more time and care to collect the facts than the writers of informal letters. They sometimes could also rely on official information. The Spanish monarchy, whose domains included Naples, Sicily, and the new world, had a long experience in coping with earthquakes.[38] After a quake, the authorities sent commissioners to the afflicted regions to survey the destruction and assist in reconstruction. On their return, they submitted an official report. Some Italian authors clearly had access to this information.

The physician Giorgio Baglivi, who wrote on the Italian earthquakes of 1703, referred to various official reports. He gave a list of places in the Kingdom of Naples with the number of victims and a note on the state of destruction, "taken from a Neapolitan diary, on the order of the viceroy printed by the bookseller Bulifonio", and a similar list from the Papal States, taken from a report by the apostolic commissioner.[39] The report from the viceregal auditor Alfonso Uria de Llanos on the province of Aquila was also printed separately as a four-page pamphlet in Rome.[40] The letters by the apostolic commissioner, Pietro de Carolis, were printed in a compilation of materials on the earthquake.[41]

The great earthquake of Lima (Peru) of 1746 was described in a Spanish report, written by order of the viceroy.[42] The title of a printed report on an earthquake in Valencia in 1748 states that it is drawn from the testimonies sent by the judges and the governors to the governor general of the Kingdom; at the end, the author confirms with his name that the text conforms to those testimonies.[43] The Italian author Vincenzo Magnati, writing on an earthquake in the West Indies in 1687, even included a report in the original Spanish.[44]

Although they relied on the same kind of sources, there are some marked differences between pamphlets and histories. Pamphlets had to be marketed as soon after the event as possible, when the impression was still fresh and curiosity most intense. A pamphlet on an earthquake in central France at 26 January 1579 has a privilege dated January 29. The first report on the big earthquake at Lima in 1746 was reportedly published (in Lima) within four days after the quake.[45]

Of course, commercial considerations played a role in book publications as well. Some authors freely admitted that they had written their work at the request of a publisher. The mathematician Johann Rasch had just published a book on comets when in 1582 Vienna was struck by a minor earthquake. The publisher thereupon asked him to write something on earthquakes as well. Rasch complied by producing a compilation of some earlier treatises on earthquakes.[46] John Ray was preparing his *Three physico-theological discourses* for the press, including a description of the earthquake of Jamaica, when England itself was struck by an earthquake. He decided to add something on this latter earthquake as well, partly because of this coincidence and "partly at the Request of the Bookseller".[47]

Full histories, however, might appear years after the events they described. Their authors therefore had more opportunity to take a more detached view of the events and describe them on a more general level. Agatia Di Somma wrote on the earthquake of Calabria of 1638. He had witnessed the quake himself, but his own experiences are only mentioned in passing. His book is not a pamphlet, but a history of nearly two hundred pages, published in 1641, three years after the quake. In the intermediate period, Di Somma had spent much time collecting materials and documenting the effects of the earthquake in the various places in Calabria. He also included some events that had happened after the big quake of 1638.

As we saw, printed pamphlets contained more than just factual information. This was still more true for the larger histories and here too, new editions often made additions or adaptations of their own. A serious history required first of all an appropriate style. The description of the Sicilian earthquake of 1693 by Alessandro Burgos, mentioned above, was in itself rather factual and apparently written in some haste, but it clearly distinguished itself from the ephemeral genre of pamphlets by its flowery rhetoric. When it was translated into German, its Augsburg publisher added an engraved frontispiece and a number of fold-out plates, well executed: maps of Italy, Sicily, Palermo, six views of Sicilian cities, and one view of the Streets of Messina.[48]

Another German history of this same earthquake, this one published at Nuremberg, is no less than 140 pages. Most of these, however, are filled with a description of Sicily, its cities and monuments, and a history of Mount Etna. Some six pages are devoted to earthquakes in general. The story of the earthquake itself, mainly a list of the destructions in the various places of Sicily, is less than thirty pages. The lack of real information is compensated by flowery rhetoric. It looks as if the publisher wanted to take advantage of the interest the earthquake had elicited, without really having much to say.[49] Such geographical or historical elaborations are much rarer in pamphlets. Only by the eighteenth century, some pamphlets follow a similar template.[50]

The Italian author Vincenzo Magnati wrote a book on the occasion of the earthquake in the Kingdom of Naples of 1688 of over four hundred pages. The first two hundred pages are devoted to general considerations – the religious meaning of earthquakes, earthquakes in ancient authors, a chronicle of earthquakes in the seventeenth century. Only in the second half does he get to the 1688 earthquake. He listed the damage and the other events in all the major towns, but with any region or city that he mentions, he first gives an elaborate description of its ancient history.[51] The Umbrian earthquake of 1703 was the subject of a history by Francesco Angelo Grimaldi. Nearly one half of his text consists of a chronicle of earthquakes from the Creation to the year 1703. Thirty-eight pages are devoted to the description of the earthquake proper.

Many histories of earthquakes were translated. After Salvatore Ruffo published an Italian history of the earthquake of Palermo in 1726, a Leipzig publisher published both a Latin and a German translation. Sometimes, such translations were elaborated further. The official Spanish report on the great earthquake of Lima (Peru) of 1746, printed at Lima, was translated into English and formed the core of a separate book, published in London in 1748. The book is nearly 350 pages long, but less than seventy make for a translation of the original Spanish text. The remainder consists of a general description of Lima and Peru, and some appendices on the earthquake of Jamaica which had happened half a century earlier, in 1692. The English book was then again translated into French; this translation was published at The Hague in 1754. The translator added still another part, some physical considerations on the earthquake by Stephen Hales, originally presented to the Royal Society in 1750.[52]

What in the end would the reading public learn from this new flow of information? The major (implicit) lesson appears to have been that earthquakes were common phenomena of nature that occurred not just in antiquity or in distant and exotic lands, but in many more familiar places as well. As to more general knowledge of earthquakes, many of these works contained some summary of existing theory, mostly following the points mentioned in the previous chapter. On the other hand, the actual description of earthquakes pays only little attention to physical aspects. The facts contained in such reports are mostly of two kinds: damage to prominent buildings, especially churches; and things that have happened to individual persons.

The authors' interest in individual stories concerns in the first place prominent persons, such as princes and bishops. However, stories about common people could be included as well, if what they went through was spectacular enough. There is a number of stock stories that turn up in description after description, sometimes up to the present time: small children who are saved; a child found alive at the breast of his or her dead mother; and so on.[53] There is little doubt that small children being saved made a big impression on people (as they still do). So, when the physician Johann Rudolf Bullinger visited Yvorne shortly after the place had been buried by a collapsing mountain, he made a point of visiting two children, one of seven weeks, the other six years old, who had been dug out alive.[54]

The fact that earthquake histories, however meticulously researched, in the end mostly went back to personal stories, makes them rather predictable. Such stories probably circulated orally before they were picked up by historians. They were therefore filtered and shaped by the rules of oral transmission, although it is clear that they served religious ends as well. The theologian Johannes Heidenreich commented on a story of a butcher who was saved while his whole house collapsed: "from this there is for pious souls a clear lesson to draw, that these things happen under the providence and guidance of the omnipotent God".[55]

Not everybody was interested in the details of the events. Authors of sermons, edifying treatises or works of literature did not have to bother too much about their sources. In many cases, they hardly wrote about the event itself. What counted was the interpretation. Philosophers too generally based their interpretations more on ancient authorities than on investigations of what happened. Incidentally, their curiosity might bring them to collect more information.

Ideas on earthquakes were in large part determined by the available sources and who controlled the flow of information. In the course of the seventeenth century, new groups of intellectuals came to the fore, new channels of information were opened, and new textual media became available. This naturally changed the way how phenomena were described and perceived. Some of these developments will be the subject of later chapters.

Notes

1 The classical work on the printing press is Eisenstein (1980). Recognition of the importance of the rise of postal services is more recent. See Behringer (2006); Caplan (2016); Pettegree (2016) esp. 17–57, 167–181.
2 Keckermann (1607), dedication.
3 Scheuchzer (1706–1708) I, 124.
4 Günther (1890) 241, 252.
5 Evelyn (1955) 115.
6 One text on the flyleaf of a Bible is printed by Overvoorde (1907); another is given (facsimile and transcription) by Houtgast (1991) 46. The copy of Descartes' book is in Utrecht, university library, Y qu. 229.
7 E.g. Maior (1591); Heidenreich (1597); Liebergen (1692); van Bergen (1692); ten Kaate (1692); Stillingfleet (1750); cf. also Llano y Zapata (1748) 8, 26.

8 Schmidt (1976–1977) with examples from (a.o.) 1692, 1693, and 1755. Condorelli (2013) 154–155.
9 Nicolosi (1982) lists nine such inscriptions from the city of Catania commemorating the earthquake of 1693.
10 Clusius to Hugo Blotius, 1590 Nov. 15 st. vet. Vienna, Österreichische Nationalbibliothek. Clusius correspondence CLU-C262.
11 On pamphlets, see a.o. Schwegler (2002) 15–32; Pettegree (2014).
12 *Erschröckliche Newe zeytung* (1542): "Das hab ich euch in eil müssen anzeigen. Die post will weck."
13 Gross (1618), quote on p. 3.
14 de Poardi (1627, German version).
15 Condorelli (2013) 142 describes how the news of the Sicilian earthquake of 1693 was divulged over a network of learned letter-writers; see also 146–148.
16 Wick (1975). See 184–185 for the text of the letter by Held; see introduction, 13–16, on the sources for his collection. See also Pettegree (2014) 89–91.
17 *Discours d'une ... copie du grand deluge* (1580), [A4]v–B2.
18 *Warhafftige und erschröckliche newe Zeytung* (1590).
19 Wettersteint (1673).
20 *Narratio de gravi terrae motu in Helvetia apud Bernates.* The manuscript is preserved in Wolfenbüttel, Herzog August Bibliothek: 64.24 extravag. fol. 159+v.
21 *Warhafftige und erbaermliche Zeittung* (1584).
22 Cf. Bogucka (1999) 313–317 (for earthquakes see esp. 316).
23 Wassenberch (1981) 80. The same event is also reported by Trithemius (1690) II, 649 and by Ragor (1578) 56–57, referring to Nauclerus.
24 [*Bericht von einem Erdbeben bei Nizza in Italien* (1564)]. The copy in the Zentralbibliothek Zürich misses the upper part, with the illustration and the title. The name of the addressee has also been cut; I took it from the Augsburg pamphlet. On the report Weber (2015) 59–60.
25 *Ware / Erschröckenliche / und Erbermmliche Newe Zeytung* (1565).
26 *Grausame und erschreckliche geschicht und zeitung* (1564).
27 *Warhafftiger Bericht* (1564). Errors in copying Roman numerals are of course easily made. See also Schmidt (1994) 392.
28 *Relation* (1726): "Nach dem zu Wien gedruckten Original".
29 *Discours merveilleux* (no place, 1580), "suyvant la copie Imprimée à Pairs [sic]...". See *French vernacular books*, numbers 46767, 46768, 46769, and 46770. I was not able to check to what extent the various editions agree. A final section, "Exortation au peuple chrestien...", is lacking from a 1874 reedition of the Lyon edition, but I have not been able to inspect the original. See also Rueda and Fernández (2008) 585–587 for an analysis of the pamphlets after the 1680 Malaga earthquake.
30 *Warhafftige newe zeitung* (1603).
31 *Warhaffte Relation* (1638); *Verdadera relacion* (1638); *Einkommender bericht* (1638).
32 Weber (2015) 65–66. Weber also claims that the Jesuit scholar Athanasius Kircher incorporated a Latin version of this relation into his book *Mundus subterraneus*. This seems unlikely, as this book is from 1665. Cf. also Rueda and Fernández (2008) 585.
33 Burgos (s.a.) 1. The work was originally published in Italian: *Lettera del padre Alessandro Burgos scritta ad un suo amico, che contiene le notizie sinora avute de' danni cagionati in Sicilia da tremuoti a 9 e 11 genn. 1693*, and published in Palermo in 1693. It had a large European success. See Condorelli (2013) 150.
34 Burgos (1693) 6, 17, 18.
35 Grimaldi (1703) 18.
36 Ragor (1578) 58–59; van Helmont (1648) 100.
37 Cysat (1969) 882. Cysat was at the time in the village of Art, four hours from Lucerne.
38 Cf. Batlle (1999) 71, for actions of the secular authorities in the Middle Ages.
39 Baglivi (1704) 529–531, 531–534.
40 Uria de Llanos (1703).

41 Chracas (1704) 149–173: letters by Pietro De Carolis to cardinal Paolucci, 1703 Febr 25 and 26.
42 *True and particular account* (1748) iii, 131–199.
43 Carrasco (1748).
44 Magnati (1688) 27–42. Magnati also gives detailed information on the destruction in the various places in southern Italy in 1688.
45 Llano y Zapata (1748) 31. The work itself, by Don Victorino Montero del Aguila, captain of the guard at Lima, has not been retrieved.
46 Rasch (1582). The volume contains translations of works by Nausea, Beroaldo, Konrad of Megenberg, and Albertus Argentinensis.
47 Ray (1713) 272.
48 Burgos (1693).
49 *Less-würdige Beschreibung* (ca. 1693). On earthquakes general: 109–112. On the earthquake of January 1693: 112–140. On Etna: 50–58.
50 E.g., the first three and a half pages, out of eight, of *Relaçam* (1748) deal with a history of the island of Madeira, before moving on to the earthquake.
51 Magnati (1688) 274–410.
52 *True and particular account* (1748); *Histoire des tremblemens* (1752). See also *Individual* (1748) for a Portuguese translation.
53 See Weber (2015) 56–65 on narrative conventions in reports of earthquakes, esp. 62: anecdotic events. See also Bogucka (1999) 317 for similar anecdotes in pamphlets. Stories of a child found alive after several days with its dead mother, for instance, in Recupito as printed in Stengel (1651) II, 249; Rumphius (1675/1997) 5/51; Grimaldi (1703) 9; Llano y Zapata (1748) 4.
54 Bullinger's letter is quoted in Scheuchzer (1716) 132.
55 Heidenreich (1591) B3v.

4

EARTHQUAKES IN RENAISSANCE SCHOLARSHIP

Apart from the classical tradition as transmitted by the universities, in the late Middle Ages there arose, starting in Italy, still another way to deal with the knowledge of antiquity. Whereas scholastic philosophy had been shaped by university professors for the purpose of teaching, this new movement was dominated by intellectuals who in most cases had positions in courtly and civil life. They were critical of the scholastic approach and propagated a return to the original texts, devoid of the later commentaries and interpretations, and realised that Aristotle was not the only ancient philosopher worth studying. They often took an independent position with respect to ecclesiastical interests as well. Their ultimate goal was a restoration of ancient values and classical civilization.

These people were important scholars and their contributions in fields like philology and history have long been recognized. However, they were mainly interested in the humanities – hence, the name of Renaissance humanists by which they are commonly called. For that reason, they have for a long time been ignored in the history of the sciences, as this field used to be defined by modern disciplinary demarcations and focussed on the idea of progress in the form of tangible discoveries. Only in the last decades have people come to realize not only that humanist scholarship had important repercussions on all intellectual activity, including the sciences, but also that it makes no sense for the early modern period to study the development of what we call the sciences isolated from what happened in the humanities.

The study of earthquakes offers a case in point. The empirical investigation of earthquakes initially was a part of history, not natural philosophy. Whereas scholastic philosophers wrote textbooks wherein the phenomena were discussed on a general level, humanist scholars, on the other hand, were interested in concrete events. Their treatises often do not deal with earthquakes generally, but discuss a specific earthquake. An example from the first half of the fifteenth century is the Italian humanist Gianozzo Manetti, originally from Florence, who later in life was a councillor in Naples. There, he directly witnessed the severe earthquake of 1456. He thereupon wrote a treatise on it, probably at the request of the court. In three books, he discussed first the various theories, in particular those of the ancient philosophers; second, he gave a catalogue, based on historical research, of all the earthquakes since the creation; and third, he gave a detailed description of the damage done in Naples. The treatise is based on ancient philosophy, historical research, and direct observation. There is hardly any reference to religious interpretations.[1]

The Neapolitan philosopher Simone Porzio wrote a short treatise, dedicated to the viceroy, on the earthquake, the new mountain and the opening of the earth at Pozzuoli in 1538. This was a widely discussed event, mentioned in many later works. Porzio gave a natural explanation along the lines of Aristotelian meteorology. As to the question what the event portended, Porzio referred to natural causes only. It just announced dryness, which of course might cause other harms, like bad harvests. He clearly rejected more ominous interpretations. He wrote his treatise "lest the soothsayers, the interpreters of dreams and the vulgar astrologers interpret the things that come forward under the guidance of nature in another way".[2]

A more ambitious work on the same event was written by Piero Giacomo da Toledo. This is not a true scholarly work, as it is written in the Italian vernacular. There is a dedication by the author to Piero di Toledo, marquis of Villafranqua and viceroy of Naples, which might suggest that the work was originally written in a courtly context. Besides, there is a prefatory letter "to the amateurs of the vernacular tongue" by the author Giovan Battista Pino. The work itself has the form of a dialogue between two persons, Peregrino and Suessano. The dialogue form was popular among humanistically minded scholars as it was at the same time didactic and literary.[3]

Peregrino first gives an extensive report of the events at Pozzuoli, and points out that there is much difference of opinion about their significance. Suessano then explains the causes of the phenomenon. As for the efficient cause of earthquakes, he maintains the theory of Aristotle that they are caused by winds. He does not deny that fire has emerged at Pozzuoli, but feels this cannot have been the true cause. Fire cannot persist under the earth. It must have been caused by the wind that set on fire some vein of sulphur or alum or similar material.[4]

The dialogue continues with the common elements: accompanying phenomena, signs of earthquakes, and remedies. Finally, Suessano explains the specific situation at Pozzuoli and how this led to the earthquake. He makes a point that deep wells are a remedy against earthquakes and that the new cleft will for many years protect the province. Also, he makes rather extensive use of the analogy of the earth and the human body. The caves in the earth are compared to arteries and veins. Man is a little world, a microcosm, and the trembling of the earth is compared to diseases and affects in the body.[5]

As for the meaning of the earthquake, Peregrino mentions that the "fama publica" goes that the earthquake is not just something bad in itself, but also that it is a presage of future evils. Suessano replies: "No natural thing can signify anything else than the cause from which it necessarily depends, or the effects which it necessarily produces."[6] If sometimes war or famine follow an earthquake, this is just by natural causality, from the position of the planets or the effects of exhalation. Indeed, asked in the end whether the earthquake will bring good or bad, Suessano only speaks about natural effects, on the air notably. As he explains, the common opinion has its origin in the ingratitude of the human race towards their maker. People do not understand the phenomenon and imagine that the earth knows about their misdeeds. In line with this stance, Suessano describes the final cause of earthquakes as "the good of the universe as it is known to God, great Lord of everything", without any reference to the punishment or warning of sinners.[7]

The ideas in these treatises are not based on empirical research, but on book learning. In many respects, humanist scholars followed the example set by medieval scholastics. They discuss the same set of standard items. The religious element is in many cases rather muted. They exhort to piety, but most of them do not dwell on earthquakes as divine warnings or punishments and reject the view that they announce some further disaster. If they are signs of future harm, than only by natural causality. Still, humanist scholars did not follow medieval scholarship uncritically. There are several points on which Renaissance scholars came up with new ideas. Some of these had important repercussions on later debates. So, even if their results were disappointing from a modern point of view, we have to give a short overview before we can move on to the major developments in the next chapters.

Earthquakes in moral philosophy

The new humanist scholars, with their courtly connections and place in civil life, were principally interested in moral and political questions and the study of nature was subservient to this goal. Many humanist scholars developed a special interest in the views of Plato, the neo-Platonists, and the Stoics. The Stoics were deemed especially relevant for ethics. One Stoic philosopher who became very popular was the ancient Roman philosopher Lucius Annaeus Seneca. He is of special interest for us: he wrote extensively on nature, and in that context, on earthquakes.

Seneca wrote on nature is his *Quaestiones naturales* [Natural questions]. The work is divided into seven books. Each book has a specific subject, in order: fires in the air, thunderstorms, earthly waters, the river Nile, clouds, winds, earthquakes, and comets. Book VI, on earthquakes, has the character of a separate treatise. The direct occasion seems to have been an earthquake in Campania on 5 February 63, mentioned in the very beginning of the book. The book has the stated goal of supporting the victims of the earthquake: "It is necessary to find solace for distressed people and to remove their great fear."[8] This solace is mainly sought, in Stoic vein, in the fact that earthquakes are not supernatural occurrences, but natural phenomena. The book is therefore largely devoted to an elaborate investigation into their causes. This part needs no special discussion, as Seneca's theories are on the whole very similar to Aristotle's.

Before engaging with the main topic, however, Seneca introduced some more general considerations. Commenting on the fear and trembling of the people who had escaped the quake, he argued that such a fear is unreasonable: death is equal for everybody. It makes no difference whether one is killed by a small stone or by a complete mountain, whether one expires in daylight or is buried under the earth. We are vulnerable beings who can be killed by a trifle, so there is no reason that spectacular events like earthquakes, lightning bolts, or the opening of the earth, should incite more fear than risks that look less spectacular. Earthquakes are not worse than other dangers.

Fleeing the place where a quake has happened (as many people do) makes no sense, since there is no place where one is safe for such disasters. Promising oneself good luck is idle, as everything is unstable. The fact that we are mortals, and know it, should be our solace against unreasonable fear. Specific dangers are not worse than any other death. On the contrary, since we have to die anyway, it may be felt as a privilege to be killed by a major cause. Seneca concluded this section: "If I must fall, let me fall with the world shattered, not because it is right to hope for a public disaster but because it is a great solace in dying to see that the earth, too, is mortal."[9]

In the Renaissance, Seneca's ideas on earthquakes were taken up by various authors. One of the earliest, and most influential, was the fourteenth-century humanist Francesco Petrarca, in English most often called Petrarch. Petrarch is now especially known for his Italian poetry, but in his own time he was famous for his Latin prose works, especially on moral philosophy. In his work *De remediis utriusque fortunae* [On the remedies against good and bad fortune], in two books, he advocated a moral attitude that was definitely influenced by the Stoics, although he reformulated Seneca's precepts in a Christian sense. The book would remain popular well into the seventeenth century.[10]

The work consists of a number of dialogues between two personifications, Metus (fear) and Ratio (reason). As can be expected, Reason acts as an instructor who explains to Fear that all the things that seem dangerous to us, loose their menacing appearance once one is guided by reason. The ninety-second dialogue is titled *De terraemotu* [On the earthquake]. The text follows Seneca's argument rather closely. The dialogue starts with an explanation of the dangers of earthquakes. Reason has little solace to offer. Instead of hinting at remedies, as in other cases, she can only confirm the dangers: there is no remedy against earthquakes. One cannot even prepare for them, since earthquakes strike at completely unpredictable moments. The only solace is that earthquakes are relatively rare.

The real comfort is the same as Seneca offered: If death is the worst that can happen, what difference does it make if you are killed because you are struck by a little stone, or by the Apennine mountains? In the latter case, one could even regard death as more glorious [*clarior*], since the instrument is so large. The one remedy is to arm one's soul with virtue. When even the firm earth deserts us, one can only put one's hope in God, who is immovable. In God, one is safe and will not be moved or fear earthquakes. Fear answers: "I cannot not be moved by an earthquake", whereupon Reason:

> But you can detach all hope and all desire from the earth. Do so, and you will live safely and you will keep standing when she will move or collapse. It is stupid to put a firm hope in shaky things.

This concludes the dialogue.

Many other authors on earthquakes would refer to Stoic wisdom, either implicitly or explicitly. Filippo Beroaldo, a learned physician from Bologna, wrote a treatise on the earthquake that struck his hometown on 3 January 1505, with its fore- and aftershocks. The work is dedicated to Erasmus Vitellius or Erazm Ciołek, Bishop of Płock in Poland and just like Beroaldo a well-cultured scholar. Vitellius had visited Italy on an embassy to the Pope in 1501 and visited again in 1505. On one of these occasions, he and Beroaldo may have met and concluded friendship. Beroaldo's treatise, in Latin, is clearly addressed to a learned audience, interested in humanist studies and moral philosophy, rather than just spectacular wonders. The work was published, together with a treatise on the plague, in Bologna in the same year and republished posthumously in 1510 in Strasbourg.[11] As such, it was the first substantial work on an earthquake to be printed and it is worthwhile to have a closer look at it.

Beroaldo's approach is principally a moral one. The description of the earthquake itself is preceded by a long digression on the misery of human life. Nature is not so much a mother as a stepmother, who hands us bitter instead of sweet. This is followed by an elaborate description of the Bologna earthquake and its various effects, with interesting personal details. Beroaldo showed himself rather sceptical of received opinions. He noticed that this severe earthquake happened in winter, whereas the received theory has it that earthquakes occur in fall and spring. He also noticed that vaulted buildings, unlike the opinion of Pliny, did not prove safer than others. So, everything is set to explain earthquakes as the unpredictable and unavoidable evil that Seneca described.

Beroaldo recounted that many people who for fear of new quakes preferred to sleep outdoors that January, became ill from the cold and humidity. He commented that apparently these persons did not know that one cannot protect oneself against earthquakes. In a show of Stoic *apatheia*, Beroaldo himself wanted to keep using his bedroom, but he gave in to his wife who wanted to move down to the first floor, where the quakes were felt less.[12] (Beroaldo probably felt this to be a bad idea, because of the exhalations which were believed to be released during an earthquake and generate plague. The physician Grataroli explicitly warned that in order to avoid these toxic vapours, one should sleep at high places, not near the ground.[13])

Thereupon, Beroaldo passed in review the various opinions on the causes of earthquakes, stating at the very start that there is disagreement among the philosophers and that nothing can be said with certainty. He considered Aristotle and Albertus Magnus as the main authorities. He also mentioned the astrological explanation and the opinion of Strato, who felt that earthquakes were caused by the conflict between heat and cold. Thereupon, he came to the theological explanation:

> the religious hold it that the wrath of God is the cause of all these ills, who hurls this kind of darts in order to punish sins and sinners. And it is certainly profitable for life and expedient for mortals to believe that such torments are inflicted and imposed by an irate God, so that they will behave in a more inoffensive, pure, and modest way. This is asserted by the founders of our Church....
> (Follow some biblical sentences)

However, although the theological explanation might be useful, Beroaldo in the end kept aloof and preferred a sceptical stance:

> I am of the opinion that this evil, as several of the same kind, lies hidden in the majesty of nature and the secret council-chambers, and even if to a certain point they are known, they can never be predicted, however much the sharpness of the minds or the skill of the ingenious have tried, in long nightly studies, to penetrate the mysteries of truth, which are locked away in the inner sanctuary. For never do we have such a familiarity with God, that he wants us to take part in all the secrets....[14]

He was more explicit on the question whether earthquakes served as portents that announced disaster, a view he decidedly rejected. Actually, Pliny's saying that no earthquake has struck Rome without portending something evil, roused his indignation:

> Are such the celestial wraths? Is so strong the desire to rage against miserable humans, that an earthquake, whereas nothing more terrible exists, would not be enough in itself, if it would not be also a prodigy and if not by this evil some other evil would be announced?[15]

Besides, there are the standard elements of a scholarly overview. Beroaldo listed a number of earthquakes mentioned by ancient authors, he gave a categorization of earthquakes, and he discussed signs and effects. Earthquakes bring other evils, in particular plague because of poisoned air. He dismissed the Hermetic view that earthquakes can be predicted from the signs of the Zodiac. As for remedies,

Beroaldo clearly set no great store in the traditional recommendations. The best protection is to go living far north. In a nod of politeness to his dedicatee, he admits that Poland seems to be free of earthquakes. However, the one and only sure remedy is not to fear death.

> Who despises death and treads the fear of death under foot, not only is safe when looking upon lands being shaken and torn from their adjacent parts by a quake, but also upon his own house being swallowed by waves or devoured by clefts.[16]

This might have seemed a nice moral lesson to conclude. However, under the circumstances, it appeared to leave Beroaldo unsatisfied and he continued with some qualifications. He pointed out that lack of fear depends on one's constitution. In northern regions (as those were Ciołek lived), people get a robust constitution and despise fear and danger more than southerners. However, even in robust people, there is a natural inclination to be startled at sudden danger. The Stoic ideal of "apatheia", the complete overcoming of passions, is out of reach for humans. Instead of with an uplifting moral lesson, Beroaldo concluded with the story of one of his fellow citizens who was so upset by the earthquake that he completely lost his mind and in the end committed suicide.[17]

Humanist historians definitely attached a moral meaning to earthquakes, in the sense that earthquakes demonstrate the vanity and instability of all earthly things, but were reluctant to interpret them as divine punishments. They squarely dismissed the view that they should be seen as omens that announce further evils. In the course of the sixteenth century, this approach would increasingly lose ground to the confessionalized views propagated by the Churches. However, it would not be completely forgotten. Petrarch's dialogue remained well known. Giovanni Battista Della Porta, even though he wrote a meteorology rather than a moral treatise, quoted it at length.[18] Di Somma's history of the Calabrian earthquake of 1638 clearly situates itself in the Stoic tradition. There is no reference to Seneca, only a passing remark that the author had spent a year in a small village with no other company than the ancient Stoic philosophers,[19] but his whole history serves to highlight the way that fate is playing with humans and to demonstrate the vanity of human efforts. Among other things, Di Somma seems to delight in examples of people who, misguided by piety, sought refuge in church buildings or other dangerous places and were killed when these collapsed.

Physicians and subterranean fire

Although most humanist scholars of this period were not primarily interested in the explanation of natural phenomena, there are exceptions. In particular, physicians had a professional interest in the workings of nature. With the growing dissatisfaction with Aristotelian philosophy, the accepted interpretation of nature came open for criticism, and many physician-philosophers jumped into the field.

One of the topics that were debated was the constitution of the earth and the nature of the subterranean *spiritus* that made the earth quake.

There were two main factors that seem to have stimulated new ideas on this point. In the first place, in the Middle Ages some natural waters had acquired a reputation for their healing qualities and a number of spas had arisen. Physicians took an interest in these phenomena as they were eager to bring them under their supervision. As a consequence, healing waters became the subject of learned study and natural history, initially outside the universities. The growing literature on healing waters made people aware of the distribution and variety of such springs in Europe. Hot springs appeared to be actually quite common and their presence required an explanation. Here again, the growth of information, caused by better communications, changed the field of knowledge.

Medieval philosophers had largely ignored hot springs, as Aristotle had not discussed them. By the end of the sixteenth century, however, they had become standard topics in works on meteorology. As to their origin, a variety of ideas was suggested, but most physicians and philosophers accepted the existence of some form of subterranean fire. This did not mean that the earth had a fiery nature or that it was all hot inside. (In Aristotelian philosophy earth as an element was cold and dry.) These fires were supposed to be of an accidental character and burn locally. Of course, this theory gave rise to a host of new questions, such as: what kindled these fires, what fuel made them possible, and how was it possible that they remained burning for such a long time?[20]

The second factor was the invention of gunpowder. Already in the fourteenth century, the Italian physician Giovanni De Dondi compared the eruption of subterranean smokes from the earth with the working of bombards (primitive cannons).[21] A much more powerful metaphor was offered by the invention of siege mines by Italian engineers at the beginnings of the sixteenth century. The technique involved digging a tunnel that went under the ramparts of some besieged fortress, to put there a large quantity of gunpowder, and make it explode. The explosion would make a large breach in the wall overneath, creating an access for the besiegers. This showed the potential power of subterranean fires and suggested that naturally occurring substances might be able to do the same. At a time that philosophers were wondering about the nature of the subterranean fire, siege mines therefore offered a model for how it worked.[22]

According to the accepted theory, dry exhalations were not just responsible for earthquakes, wind, and lightning, but also for all kinds of fiery meteors. The latter ones happened when exhalations escaped into the atmosphere and caught fire in the air above the earth. So, the exhalations were evidently inflammable and could easily be made responsible for subterranean fires as well. All these phenomena were thereby integrated into a coherent theory. Many authors came to see the dry exhalations of Aristotelian meteorology as some explosive substance of a sulphurous or nitric character.[23]

In proposing such ideas, authors proudly and self-consciously proclaimed their independence from traditional Aristotelianism. In practice, however, their ideas are for us hard to distinguish from those of Aristotle. The main difference seems to be that Aristotle spoke of winds that "turn back into the earth", whereas the new ideas spoke of spiritus that were generated within the earth itself. Later philosophers often used the term *halitus*, "breath", instead of exhalations, maybe because of this difference.[24] But since the dry exhalation of Aristotle, which caused the wind, also was generated by the earth, the difference is often rather subtle.

Andrea Cesalpino, professor of medicine at the University of Pisa, may serve as an example. He continued the peripatetic tradition, but in a critical vein. Religious questions are completely absent from his work. Nowadays he is especially known for his studies on the blood flow and on botany, but in his work "Peripatetic questions" (1569), he discussed such subjects as the tides and the saltiness of the sea, both subjects traditionally dealt with in meteorology. He accepted the existence of subterranean heat, but had difficulty with existing explanations. He felt it hard to believe that the influence of the heavens could be felt deep within the earth and instead speculated that the sun might heat the sea water; the warm water then penetrated and thereby heated the earth. The heat caused the earth to produce exhalations, which might then be turned into fire. The fire caused all kinds of phenomena, including earthquakes.[25]

Another physician, Antoine Mizauld, in a work first published in 1554, rejected the view that earthquakes are caused by wind enclosed in caves against its nature. However, referring to pristine philosophers like Hermes and to the knowledge of the common people [*rusticus*], he still felt that earthquakes are caused by "spiritus" generated by the earth. The same spirits are also responsible for wind and thunder and lightning. He compared their effect on the earth not so much with siege mines but with chestnuts that are being roasted. He still listed all the standard points, normally with standard answers, although his classification of earthquakes is somewhat different; in particular, he was sceptical about the Aristotelian distinction of tremor and pulsatio, "whatever this is".[26]

Probably the most influential author on the subject of subterranean fires was Georg Agricola. He had studied at Italian universities, where he had absorbed the culture of humanist scholarship. As a physician, Agricola then worked mostly in the German mining regions and he used this position to make a thorough study of the subterranean world. He is best known for his work on mining (*De re metallica*), but he published a number of other works on the topic. For our subject, the most relevant are *De ortu et causis subterraneorum* [On the origin and causes of subteranean things] and *De natura eorum qui effluunt e terra* [On the nature of things that flow from the earth].

Agricola did include sections on earthquakes, but he had little personal experience of these and his argument is mainly a discussion of the works of ancient authors.[27] (Actually, he seems to have been used as a source for such references by later authors.) That is not to say that his text was just a simple compilation. As stated before, he came up with his own classification. He also criticized Aristotle's opinion that earthquakes were more prevalent at specific times of the year or the day.[28] He was more familiar with hot springs and these were discussed at considerable greater length. Agricola gave an elaborate and thoughtful discussion of the various possible explanations. He agreed that some form of subterranean fire clearly existed, but he disagreed with most of his predecessors as to its fuel. Rather than sulphur or saltpetre, Agricola concluded that the heat was generated by the burning of bitumen, a compact fuel that was able to burn even in contact with water. His considerations were quite influential and were referred to in many later works. For instance, Cesalpino in his work on metals now discussed burning bitumen as a source of subterranean heat.[29]

An author who more specifically linked the theory of subterranean fire to earthquakes is Hieronymus Cardano. He was best known as a physician and astrologer, but was also active as a mathematician, inventor, and all-round scholar. As a learned physician, he studied the natural world and tried to solve its mysteries. His most important work from a natural philosophical perspective is *De subtilitate* [On subtlety], a large encyclopaedic overview of the world. Leftovers from the writing of this book were later published in *De rerum varietate* [On the variety of things]. Both works were widely read and saw many reeditions.

Earthquakes are especially discussed in book II of *De subtilitate*, which is devoted to the elements. In his discussion of the air, Cardano first explained that stagnant air putrefies and becomes poisonous. He then rather abruptly moved to earthquakes. Like Seneca, he distinguished three types: *inclinatio*, *succussio*, and *vibratio*; the first is the most dangerous, the latter the least. Cardano continued with a digression on the wonders that often accompany earthquakes: lakes and wells originate, the flow of rivers is reversed, and horrible sounds are heard, "as voices of those who fall in battle". Prodigious as these things seem in themselves, they all depend on the quaking of the earth. "And maybe this cannot happen without prodigy, though it is certain that they happen by natural causes".[30]

As for causes, Cardano felt that they lay in matter which was burning inside the earth: sulphur, saltpetre, or bitumen. The spirits that escaped from the fire needed an exit. If they did not find it, they forced themselves a way out, causing an earthquake. So, like other philosophers, Cardano gave the spirits from burning matter the role earlier played by Aristotelian exhalations. He elaborated upon the qualities of the respective combustibles and their various effects in earthquakes.[31] Cardano discussed several other aspects, some of them familiar from scholastic discussions. He discussed how wells might show signs of imminent earthquakes. He explained why some regions, like Egypt, were not prone to earthquakes. This concerned above all regions that were solid and could not take in air; regions that were loamy and had no cracks; and sandy regions that let the air perspire. Cardano included the caveat that the earth might have a different constitution at greater depths than at the surface.[32] He also included an explanation of the origin of mountains, a topic not discussed in traditional meteorology. He distinguished three different causes: some mountains were inflated by repeated earthquakes, like a bladder; some were blown together by the wind, as often happened in Africa (most likely Cardano was thinking of the sand-dunes in the Sahara desert); some originated from stones left by the sea.[33]

Systematics and coherence were not Cardano's strongest point. In still another work, on eternal secrets, in a chapter on "the seven calamities of the human kind", he referred to an event that we would classify as a volcanic eruption. In this case, he denied that the fire originated in the earth but instead maintained that it came from burning salt. Since seawater is salt, conflagrations in or near the sea are bigger and last longer.[34]

Impact on traditional learning

The new humanist scholarship could not fail to affect the traditional teaching of meteorology and of philosophy generally, be it often in a superficial way. Some authors simply put traditional scholastic knowledge in a more literary form, to make it palatable for a courtly audience. Many would include some of the more recent theories, like those by Agricola, in their commentaries, while keeping an overall traditional framework.

A commentary on Aristotle's meteorology by Francesco Vicomercato, first published in 1556, explains that the external wind does not contribute much to earthquakes. They are caused by wind that is generated within the earth "by the force of heat that is inherent there [*insitus*], especially in places where there is found some form of fire, kindled within the earth, as commonly happens in sulphurous and bituminous places".[35] This view is based on contemporary humanist scholarship rather than on Aristotle.

A small treatise on earthquakes was appended to a treatise on the wind, as a related subject, by the physician Fabrizio Padovani. The two treatises were published posthumously in 1601 by the author's son with a dedication to the duke of Urbino. The treatise can be regarded as humanist in its extensive use of classical sources, but for the rest it is utterly unoriginal. It mainly passes in review various scholarly opinions on the standard scholastic points (causes, effects, classification, signs). The one thing that is remarkable is that Padovani refuses to discuss the final causes: earthquakes "do not happen from nature's intent, rather, they spring from mere accidents. Therefore an earthquake does not have a goal and for the same reason it has no use and does not bring any good. So far on causes".[36]

Given the penchant for astrology among sixteenth-century scholars, one would expect that astrological explanations figure prominently, but this appears hardly the case. Astrological explanations certainly existed. Among the standard opinions on the cause of earthquakes that were rehearsed by scholastic authors is the idea attributed to the ancient Babylonians that they are caused by the stars. Typically, this idea is rejected. Also, there existed a (pseudo) Orphic text (sometimes attributed to Hermes Trismegistos), "Peri Seismon" (on earthquakes). This text, in the form of a poem, gave keys for the meaning of earthquakes based on the position of the sun in the Zodiac. For instance, when the sun was in Virgo, an earthquake happening at night predicted plague, and at day, a bad harvest. The text saw a Latin edition in Paris in 1586. A new translation appeared in

Göttingen in 1691. It appears that this text had little influence on the learned discourse. One author, in 1580, refers to it cursorily as "dreams".[37]

One of the few philosophers who takes astrological explanations of earthquakes seriously is the Italian physician Agostino Nipho (or Suessanus, after his place of birth), one of the most respected traditional philosophers of the period. Nipho's work on meteorology is a traditional commentary on Aristotle, but with some innovative points. As to the influence of the stars, he is not referring to Hermes or the Babylonians, but sticks to Aristotle. Aristotle's text hints at a connection between earthquakes and lunar eclipses. Nipho felt he really meant celestial influences in general,

> which Aristotle gives to understand by the example of eclipses. For there occur many other stellar configurations that have both the power to generate exhalations in the bowels of the earth, and to repel those exhalations that have been raised above the earth. Aristotle did not explain these things because this is not physics, but astronomy.[38]

Apart from the formal, material, and efficient causes of earthquakes, Nipho also discussed their final causes.

> The final cause is the same as that of the winds, to wit, of the generation itself and of what is generated. The aim of the generation is the very motion by which the earth is moved; the aim of the generated thing is the good of the universe.[39]

There is no hint that earthquakes might announce any future evils. Like the medieval scholastics, Nipho was rather sceptical about perceived miracles. Phenomena such as blood rains seem miracles to the common people, but the wise man knows their causes. Still, divine miracles were not excluded. In particular, the sounds that were sometimes heard in the air were miraculous, either directly or indirectly worked by God.[40]

Attempts at a new philosophy of nature

Apart from these piecemeal adaptations, for some the criticism of Aristotelian philosophy opened up space to come up with completely different theories. Some philosophers aimed at a complete overhaul of existing knowledge, although it should be admitted that their actual results were seldom in keeping with these far-reaching claims. In a highly eclectic way, they drew inspiration from their readings of a host of ancient philosophers, or whom they regarded as such – Stoicism, (neo-)Platonism (including Hermeticism), pre-Socratics. For the moment, most of their work was based on textual study rather than empirical research. In their attempts to come up with a new explanation of nature, they also included the field of meteorology and therefore, earthquakes.

Among such authors, we might first of all mention Cardano. His immediate impact on our field was mostly because of his theory of subterranean fire, which we already discussed, but his ideas went much further. He attempted to come up with a full-blown philosophy of nature. He was especially interested in occult properties. In his view, the world was a web of correspondences and nature constantly showed prodigies and signs to warn the sage of things to come.[41] So, unlike the authors discussed above, he did regard earthquakes as signs of future events. In *De rerum varietate*, especially in its book on divination, he explained this in some detail. Large earthquakes announce war, plague, or oppression, and cause famine. If an island emerges in a river, this need not be prodigious, as this happens fairly often. An island emerging in the sea, however, announces the origin of a new empire and new laws, although not necessarily immediately – after all, the island too grows only over time. Likewise, the emergence of mountains (like the 1538 Monte Nuove) refers to princes, as they rise above their surroundings. However, because of their long duration mountains do not refer to one particular prince, but rather indicate a new dynasty. Infertile mountains refer to tyrants, fertile ones to lenient rulers, rocky and stony mountains indicate a harsh government, hills a mild and mixed government. "But it is stupid to believe that great effects can happen without great causes. And if the causes are great, there will be produced great effects from them among humans." All this is due to some form of natural causality. Cardano emphatically denied any role for the demonic.[42] In a chapter on "the seven calamities of the human kind" of still another work, on the secrets of eternity, he hinted at an astrological explanation. Referring to four major earthquakes from classical antiquity, he wrote: "They often originate in unfavourable constellations of Saturn and the fixed stars. At these times the apogee of Saturn moved from Scorpio into Sagittarius."[43]

In the sixteenth century, the explanation of nature moved from the medical profession to a wider group of lay philosophers, mainly in Italy. The Italian Bernardino Telesio is one of the best known of these humanist natural philosophers. In his major work, he came up with some new principles, which provoked considerable controversy, to replace those introduced by Aristotle. He also published some smaller works. One of these, published in Naples in 1570, is on meteorology: "On the things that are generated in the air, and on earthquakes". Telesio rejected the Aristotelian distinction between exhalations (hot and dry) and vapours (cold and wet). Vapours and exhalations are of the same essence, only more or less dense. On earthquakes, he had no original ideas but followed the in his time modern explanation wherein the exhalations (he calls them vapours) are generated within the earth. Also, he compares their working with that of gunpowder. In most other details he rather closely follows Aristotle, to whom he even refers. So, earthquakes are frequent in hot, dense, and cavernous lands: heat is needed to generate the vapours, the caverns can collect them, and the earth's density makes it hard for them to escape.[44]

Another natural philosopher was the Neapolitan Giovanni Battista (or Giambattista) Della Porta. Among his many works is a book on meteorology,

De aeris transmutationibus [On the transmutations of the air], first published in 1610, when Della Porta already was an old man. The book is dedicated to Prince Federico Cesi, the founder of the Accademia dei Lincei, which Della Porta that same year joined as a member. The anti-scholastic tenor is quite marked. In the preface, Della Porta explained that the foundations of meteorology are known very badly and that this is due to the fact that everybody has always followed Aristotle; people prefer to be wrong with Aristotle rather than to come up with something new.[45]

The organization of the work deviates from the scholastic tradition. There are four books. Book one is on the air and wind, the second book on rain and other "wet transmutations" of the air, book three is on fiery transmutations (comets, which Della Porta recognized as celestial phenomena, are not included), and book four on phenomena on or in the earth – seas, rivers, subterranean fire and spiritus, earthquakes, and hot springs. On all these topics, Della Porta first gives long overviews of the opinions of all kinds of (often rather obscure) authorities, and then presents his own ideas.

In Della Porta's ideas on the earth, subterranean fires take a prominent place. He follows Agricola in regarding bitumen as the fuel of the subterranean fires, although possibly mixed with sulphur. A mixture of bitumen, sulphur, and water, once kindled, will keep burning forever as long as the fuel lasts. Since these substances always emerge again [*scaturiunt de novo*], the fires will burn eternally. Earthquakes are analogous to thunder: an earthquake is a subterranean thunder, and thunder is a celestial earthquake. An earthquake is an underground explosion (more or less like a siege mine), which happens when the inflammable spirits are lit by the subterranean fire. That is why islands and other places where permanent fires are burning, are so prone to earthquakes. That earthquakes also occur at places without volcanoes or hot springs, does not disprove this point. There are subterranean canals that can bring the fire everywhere. The latter point would be taken up eagerly by later authors.[46]

Della Porta's ideas were clearly influenced by his familiarity with the Campi Phlegraei, close to Naples. He gave a detailed account of the earthquake at Pozzuoli in 1538, with the emergence of the Monte Nuovo. He had only been a child at the time of this eruption, so he had to rely on written information. He is one of the first natural philosophers who bases his ideas at least in part on the information given in contemporary pamphlets and histories.[47]

Notes

1 Heitzmann (2004). On the earthquake of 1456, see especially Figliuolo (1988).
2 Porzio (1551) 8.
3 For dialogues, cf. Dal Prete (2014) 300; Martin (2011) 64–66.
4 Toledo (1539) Cv.
5 Toledo (1539) B3–B4.
6 Toledo (1539) Dv.
7 Toledo (1539) B3; see also D2, D4.

8 Seneca (1972) 129.
9 Seneca (1972) 141.
10 I used the edition Petrarca (1649). The dialogue on the earthquake is on pp. 586–589.
11 Beroaldo (1510). See Weber (2015) 41–43.
12 Beroaldo (1510) Biij + v.
13 Grataroli (1558) 63.
14 Beroaldo (1510) [B7v].
15 Beroaldo (1510) Cij. Cf. Pliny, Nat.Hist. II, 200.
16 Beroaldo (1510) Cij v.
17 This latter part was omitted in Rasch' translation, Rasch (1582).
18 Della Porta (1614) 183–184.
19 Di Somma (1641) 143.
20 Cf. Vermij (1998).
21 Vermij (1998) 332.
22 Duffy (1979) 11–12.
23 Cf. the case of Gregorio Zuccoli, in Martin (2011) 74.
24 E.g. Piccolomini (1597); Fromond (1627).
25 Cesalpino (1588) 458–462.
26 Mizauld (1555) 130v–137v. The comparison with chestnuts, and eggs, also in Albertus de Orlamunda (1502) K1.
27 Agricola (1546, *De ortu*) 24–34; (1546, *De natura*) 150–156. A manuscript "de terrae motu, 1544" was among Agricola's papers when he died, but appears to have been lost. See Prescher and Wagenbreth (1994) 211 (number 9).
28 Agricola (1546, *De ortu*) 33; (1546, *De natura*) 156.
29 Agricola (1546, *De ortu*) 13–16, 34–35. Cesalpino (1602) 18–19. See also Vermij (1998) 334–337.
30 Cardano (1611) 146.
31 Cardano (1611) 146–150.
32 Cardano (1611) 150–151.
33 Cardano (1611) 151–152.
34 Cardano (1663/1963) X, 14.
35 Vicomercato (1565) 132. The section on earthquakes covers pp. 128–143. On p. 142 (on classification), he refers to Agricola.
36 Padovani (1601) 161: "non ex intentione naturae, sed potius ex accidentis; idcirco Terraemotus caret fine, & propterea nullam habet utilitatem, neque ullum assert commodum & de causis hactenus."
37 [*Peri seismou*] (1586). On the text, cf. Dagron (1981) 93; also 92–95, on earthquakes and astrology among the Byzantines. *Des causes et effects* (1580) 14.
38 Nipho (1560) 383. A later author who included astrology was Flaminio Mezzavacca, on whom below.
39 Nipho (1560) 394.
40 Nipho (1560) 189–190 (blood raims), 103, 189–190, see also 388.
41 On Cardano, see a.o. Céard (1977) 229–251.
42 Cardano (1663/1966) III, 278–279.
43 Cardano (1663/1966) X, 14–16.
44 Telesio (1570) 8 (cap. 11). On his meteorological theories, see Almagia (1961); for earthquakes in particular, 162–163.
45 Della Porta (1614) prooemium. On the anti-Aristotelian tenor of this book, see Borrelli (2008) 85–87.
46 Della Porta (1614) 182, 195–196.
47 Della Porta (1614) 197–198.

PART II

Early modern confessionalized science

5

THE ASSAULT ON NATURALISM

Whereas in Italy, in the course of the sixteenth century a literature on earthquakes developed that was based on ancient philosophers and new medical theories, and was reluctant to attribute to earthquakes any prophetic significance, north of the Alps, the development went into a different direction. That is not to say that humanist scholarship had no impact here at all – it certainly had. But the discourse came to be dominated by the growing religious sensibilities of the period. The Protestant Reformation would harness natural learning into the service of the new Churches.

The interpretation of portents

The first decades of the sixteenth century were a period of crisis and uncertainty. The controversies within the Church, the fear of the invading Ottoman armies, and other events created a climate wherein many people expected the end of the world to be very near. Signs of the impending end were everywhere. People saw all kind of images in the sky – flying dragons, armies fighting, religious symbols. Well-known in modern historiography is the panic caused by a conjunction in the sign of Pisces in 1524. It was believed that this conjunction announced a second deluge, which would inundate a major part of the earth. Such prophecies, based on conjunctions, had been a recurrent phenomenon in European culture since the twelfth century. Nor would 1524 be the last case; a conjunction in 1583 caused similar unrest. However, the panic in 1524 was especially widespread. The power of print media brought the debate to many more people than before. Historians have spoken of a "media event".[1]

The general climate also had its impact on the understanding of the natural world. Comets, monstrous births, thunderstorms, flying dragons, and so on, were not just natural occurrences, but signified either that the end of the world was approaching, or showed the activity of the devil (or both). Such phenomena obtained a lot of attention. They were commonly referred to as portents or prodigies, a term derived from ancient Roman pagan religion. As such were indicated events that appeared to deviate from the ordinary course of nature. Prodigies were not just spectacular, but they were also felt to have meaning. Labelling something a "prodigy" implied putting it in the realm of the preternatural. Another word used at the time, "monstrum", literally means: something that is shown. Traditionally, such signs foretold war, famine, plague, and the death of Kings.

Not everybody was carried away by the sense of crisis. The great humanist scholar Desiderius Erasmus appeared unimpressed by the many strange events that were reported in his day. When he did mention such reports, it was usually to poke fun at them. That is not to say that Erasmus did not care about religion, but he rejected the miracle-mongering and the magical elements in medieval Catholicism. Instead, he emphasized piety, morality, and the study of the Bible. Similar ideas were found in the most radical wing of the Reformation, as in the writings of David Joris. The French author Michel de Montaigne too was highly critical of reports of strange events, whereas the humanist Polydorus Virgilius wrote a dialogue along the lines of Cicero's *De divinatione* to refute the belief in prodigies.[2]

Other scholars took such events more seriously. At the beginning of the sixteenth century, the circle of humanists around Emperor Maximilian I looked favourably upon prodigies. They were especially impressed by an event in 1500–1503 when crosses fell from the sky and attached themselves to clothes, houses, and persons. Several authors discussed the phenomenon, both in Latin and the vernacular.[3] Prodigies were also made philosophically more respectable because of the mounting influence of neo-Platonist and Hermetic ideas. Authors like Cardano saw the world as a web of correspondences. By understanding these, man should be able to predict the future and maybe manipulate the world.

One of the earliest printed accounts of an earthquake outside Italy is an anonymous pamphlet on an earthquake in the Swiss mountains in 1511. Although in the German vernacular, the form is consciously literary and the work was clearly written by a person of some learning. It is not a simple newsletter, but comments on the event in the form of a poem. The text starts as the narration of a frightening dream, than moves on to the actual earthquake. Poems on a supposed dream were a well-known humanist form at the time. In a rather factual way, the poem relates where the earthquake struck (northern Italy and the Alpine area) and some of the damage done. There are some references to famous earthquakes from history and some verse are devoted to the physical explanation of earthquakes. The main message, however, is religious. People should abstain from sin. If we do not improve our lives, God will multiply the punishments. As the histories show, God can punish our sins, using nature as his instrument, or even working against nature. We should pray to God to turn away the evil, but the moral state of the world is definitely bad.[4]

Unlike medieval treatises, the poem is not describing the earthquake as an *exemplum*. Rather, it is concerned with how we should regard earthquakes in general. The emphasis is on pious and moral behaviour. In this respect, the poem may well be influenced by the thought of humanists like Erasmus. The poem dates from before the Protestant Reformation (it even invokes St. Mary), but it urges many of the same themes that the Protestants would bring to the fore, although there are also clear differences.

Portents were not necessarily supernatural. Scholastic philosophers used to make a distinction of primary and secondary (or "nearest") causes. God was the primary cause of all phenomena, who used the secondary, natural causes to further his ends. Attributing a phenomenon to natural causes therefore was not to deny that it was a manifestation of God's will. It also did not necessarily imply determinism. Obviously there was some awareness of causal chains, but the concept of laws of nature was still far away. Explanations were not sought in terms of a mechanical and necessary causation, but in the "nature of the things" themselves. Phenomena stood largely on their own, so they could fulfil certain goals without this affecting the whole machine of the world.

Still, there was certainly a tension between the natural and the divine. Philosophical explanations owed their origin to ancient Greek thinkers who wanted to refute anthropomorphic, mythological explanations. There can be little doubt that medieval theologians had come to accept Aristotle's physics and his doctrine of causality, exactly because this allowed them to distinguish clearly between an omnipotent and transcendent God on the one hand, and his creation on the other. Comets, earthquakes and so on had no divine character, but were natural phenomena. Any evil consequence that they seemed to have could be explained naturally. However, the era of the Reformation saw a strong backlash against this concept of a God at a distance, not in the last place with Luther himself. A new climate of anti-naturalism and religious fervour emerged. In this climate, earthquakes and similar events became the occasion for the expression of deep-rooted apocalyptic fears.

The radicals

The German Church reformer, Martin Luther, stood at the vanguard of the new ideology. Luther was strongly touched by the apocalyptic fever of his time. He experienced the world as a struggle between God and the devil, and he constantly tried to read the signs of God's will in the phenomena. To him, everything that happened was one large *exemplum*. Theories of natural causality were not helpful to come to grips with such a world. To the contrary, Luther emphasized the activity of angels and demons and the continuous providence and interference of God Himself.

As a consequence, Luther rejected Aristotelian philosophy and the scholastic tradition. He was particularly critical of Aristotelian meteorology. In 1522, he stated, referring to Aristotle's *Meteorology*:

> So, the blind man's guide Aristotle has written a whole book on the heavenly signs. He ascribes them all to nature and makes that they are no signs. These our scholars follow; so, one fool makes a world full of fools.

In another instance, Luther declared that there was no book he believed less than Aristotle's *Meteorologia*, exactly because it was based on the assumption that all things happen by natural causes.[5]

Luther's works were incredibly influential, but he himself hardly wrote on natural phenomena. In philosophy, his views would never become mainstream (which would have meant rejecting philosophy as such). Even at the University of Wittenberg, where he held sway, teaching could not do away with Aristotelianism altogether. Still, the anti-naturalist sentiment ran deep and influenced people's ideas at various levels. Outside of universities and established traditions, it found open expression. Since this is highly revealing for the sentiment of the times, I will discuss two of these thinkers, one rather a spiritualist, the other a nominal Catholic, who did discuss the "natural" world at length.

The physician Paracelsus (Theophrastus Bombastus von Hohenheim) is one of the most influential, but also most elusive thinkers of the sixteenth century. His ideas are closely connected to the early Lutheran reformation, but cannot simply be subsumed under this heading. His writings were widely divulged and his ideas, or ideas promulgated under his name, gained a large following and influence. There was never something like a Paracelsian organization. Within many Churches, however, they formed a radical, more spiritualist strain, often looked at with suspicion by official orthodoxy. Paracelsians definitely aspired after a "Christian" philosophy of nature, but refused to be incorporated into a new scholastic philosophy.

Paracelsus developed into a person of some notoriety early in his career. As the town physician of Basel, he gave lectures that drew attention for their vehement anti-Aristotelianism and anti-Galenism. His not always tactful behaviour raised much enmity and he had to leave his position in 1528. He had not published very much by 1531, however. We know of two medical treatises, one on "the wood Guaiaco", from 1529, and one on the French disease (syphilis), from 1530. However, he had also entered the field of prognostics. In 1529, he published a "Practica on Europe", which met with some success, as it saw at least five different editions in 1529–1530. Probably in 1530 he also published a prognostication (now lost) "concerning a confederation".[6]

The appearance of the great comet of 1531 then occasioned him to publish several works wherein he interpreted such prodigies as warnings in the political–religious strife of the period. In 1531, he published a pamphlet on the comet itself.[7] When in October of the same year, shortly after the appearance of the comet, an earthquake occurred in Switzerland, Paracelsus took up the pen again and published a small pamphlet: "Explanation of the earthquake that occurred after the disappearance of the comet in the Swiss mountains in 1531."[8] Three other pamphlets followed shortly thereafter, on a rainbow, another comet, and a few more signs.

Most of the messages in the earthquake pamphlet are concerned with the civil war in Switzerland between Protestant and Catholic cantons. Paracelsus regarded the earthquake completely from a biblical standpoint, that is, as a herald of the Last Judgment. He referred to the earthquakes that had happened in Germany twenty years earlier (apparently the one from the poem of 1511), which he identified as the earthquakes of which Christ speaks in the Gospels. The earthquake of October 1531, which occasioned the pamphlet, happened only in Switzerland and according to Paracelsus addressed this region in particular. It was a warning that the country should take care not to fall into a civil war. "For every earthquake is a terrestrial comet, that indicates a *bella intestina*, that is, wars where the members of the family are up against each other."[9]

Paracelsus states quite emphatically that earthquakes are not natural, but supernatural occurrences.

> Although a flower-covered philosophy might enter here and explain this earthquake in a natural way, but Christ's authority is sufficient against them all, that here is nothing natural to look for. (…) That a thing can be considered in a natural way and can be understood naturally by the imagination, does not mean that it is as the imagination sees it.

Christ was a man, but it is no use to speculate about his complexion or other human properties; what counts for us, is that he is God. "The same with these things: one can consider them naturally, but they are supernatural."[10]

Like other authors we will discuss hereafter, Paracelsus also spoke out more generally on the origin and meaning of such phenomena, but these considerations were published only after his death. Apart from the works just mentioned, Paracelsus published only a few other works during his lifetime, all concerning medicine and surgery. He tried to publish several more medical works, but was prevented because people objected to his violent tone and his anti-Aristotelianism. He nevertheless appears to have written a good deal more: works not only of medical, but also of philosophical, theological, alchemical, and magical content. After his death in 1541, these works circulated in manuscript among his followers. Starting in the 1560s, they were collected and published by various editors. Obviously, this means that one cannot always be sure of the correct reading of a text, or even of its authorship, although it should be admitted that some of the sixteenth-century editors did quite a fine job.

Paracelsus appears to have aimed at a complete overhaul of existing philosophy. It is therefore not surprising that among his philosophical works, one also finds a book on meteorology, *Liber meteororum* (or *Das Buch Meteororum*).[11] It may date from a rather early point in Paracelsus' career, but it was first printed at

Cologne in 1565 or 1566. His explanation of the various phenomena follows, on the one hand, the familiar Aristotelian pattern, and on the other hand, he applies these explanations in a completely un-Aristotelian framework, wherein nature is animated and is the domain of good and evil spirits. It is often hard to follow his thought. He did not propose a systematic "science of signs" that specifically dealt with portents, as we find in some other authors we still have to meet. Rather, in his general philosophy, the whole of nature is given supernatural meaning.

Paracelsus agreed with the traditional view that earthquakes are caused by winds. The winds blow through the air, but also through the waters of the sea, from which horrible tempests arise. Wind is so subtle, that it can pass not only through water, but also through soil and rock.

> And notice. What remains in the water and does not leave the water [sc. into the air], that comes into the earth and goes through the earth so long, until it finds an exit or is consumed by it [=the earth], which is the cause of earthquakes. (Not just of earthquakes, but of various other things as well.) For in this quality, the wind makes earthquakes, as [it makes] waves in the water, when something is standing on it.

The idea seems to be that earthquakes and the waves of the sea are analogous phenomena, both caused by winds that are blowing below the surface.

> In a like way, the wind, through the pores of the earth, small and large, in the way of a cataract or something similar, may move the earth as well, but not so easily as the water, which is more willing and puts up less resistance.

The earth itself does not generate wind, but all subterranean winds enter from the outside. Notice that Paracelsus appears not to be influenced by then current ideas on subterranean fires.[12]

The *Liber meteororum* appears to have been conceived as an ordered treatise, but besides, Paracelsus wrote several more notes on meteorology, some of which were printed in some form or another after his death and thereby preserved. There is one note specifically on earthquakes. Herein, Paracelsus explained that some earthquakes may be caused by rocks and mountains that sink (*sich setzen*), or by the collapse of parts of the earth into abysses. On the other hand, these collapses cause wind that in its turn makes the earth move.[13]

Besides many followers, Paracelsus also had many detractors and his ideas were the subject of conflicts and controversies. Many people regarded him as a dangerous revolutionary who was to upset the established order. As a consequence, his ideas were important not just for their direct influence, but also because they forced people to reformulate their own.

An equally radically anti-naturalistic, but completely opposite idea was defended, some decades later, by the French lawyer and philosopher Jean Bodin. In certain respects, Bodin embodied the new absolutist and confessional ideals. He is recognized as one of the founders of the theory of royal absolutism. He is also rather notorious for being one of the theoreticians of the early modern witch-hunt. On the other hand, he shows no signs of baroque or Counter-Reformation piety. His religious ideas appear to have been fairly irenical and during the French wars of religion, he favoured some form of toleration of religious dissidents. Although he certainly believed in the "God of nature", in demons, and in prophecy, there are no indications that he believed in Christ or the Bible.[14]

At the end of his life, he wrote a work on natural philosophy, *Universae naturae theatrum* [Theatre of the whole of nature]. It is a general overview, apparently intended for a lay audience. The work has the form of a dialogue between two figures. Theorus (Theodorus in later editions), the student, asks questions, and is answered by Mystagogus. The book allots ample space to the supernatural. Like Luther, Bodin feels definitely that not everything that happens in the world has a natural cause. So, when they start talking about the wind, it is stated from the outset that there are two kinds of wind: natural winds, excited by nature, and forced [*violentus*] winds, "that are raised outside the ordinary tenor and order of nature by the force and power of demons, or by fleeing a vacuum". It is against the nature of the wind to do violence. Each region has its own winds that are adapted to it. This is proof that good spirits [*bonos genios*] moderate the air in specific regions. In the same way, there are evil spirits that perturb the air. That is why certain regions are more often subject to storms than others. Bodin stated that demons are a legitimate subject in natural philosophy, because they are corporeal. This, he claimed, was not just the common opinion of all theologians and philosophers, but could also be proven by necessary reasons.[15]

From the wind, the discussion naturally turns to earthquakes. Here too, Bodin felt that the cause is demons. Against the argument that demons hardly can be so powerful that they can overturn whole cities, Bodin stated that it is still much less credible that this is done by the wind in a natural way. He devoted a complete section to refuting the by now common Renaissance theory of earthquakes as a kind of siege mines, originating from exhalations from burning material. He even denied the existence of subterranean fires. In hot springs, heat occurs because the waters are heated by violent motions from cataracts and the like. The smell of sulphur is clearly an indication of the presence of demons.[16]

Partly, Bodin's arguments against the theory of burning exhalations are simply common sense. As he points out, in order to have such an explosive effect, the air would have to be restricted in a very narrow space without even the smallest exit. How could the air possibly be pressed in such a way? In mountains, the soil is spongy, so that air can escape; where soil is solid, air cannot enter. Any gunpowder-like explosion should result in enormous clouds, whereas to the contrary a very pure, thin, and tranquil air is observed in earthquakes. Moreover, there are very many different kinds of earthquakes, in all kinds of different places, in different seasons. They often last a long time.

Bodin offered many examples of earthquakes which according to him could not have happened in a natural way. Some are taken from ancient historians such as Flavius Josephus or Julius Obsequens, others are more recent, as the earthquake of Ferrara of 1570. The question that remains is whether the actions of demons are natural. Bodin pointed out that the divine and human actions are voluntary, the actions of demons however are limited [*cohiberi*] by the power of God, so that they cannot do anything without command. "Hence, neither tornadoes, nor thunderstorms, nor epidemics, nor collapses [*labes*] of the earth, have ordinary causes and effects that are congruent with nature." Smoky, thick exhalations are an instrument of demons. Will-o'-the-wisps and similar phenomena are likewise a play of demons.[17]

That storms were raised by witches seems to have been an old popular belief. Theologians interpreted witchcraft as originating from a pact with the devil, hence it was a small step to attribute storms to demons; and given that philosophers felt that wind, thunderstorms, and earthquakes were related, it would seem only logical to make them responsible for earthquakes as well. However, although some theological authors might attribute earthquakes to demons, it was never an idea that gained much of a hold in philosophy.[18] Philosophers had to find other ways to explain the religious meaning of earthquakes.

Notes

1 On the events of 1524, see Zambelli (1986); Barnes (2016) 89–130. On the earlier prognostications: Weltecke (2003) 179–212.
2 Laureys (2005); Céard (2008) 163–170 (on Polydorus), 387–434 (on Montaigne).
3 There is no modern study of the phenomenon. One can find an extensive documentation, however, on Wikipedia.de, s.v. Kreuzwunder. A major source of the event is a letter by the bishop of Liège to King Maximilian, printed in Nauclerus (1614) 1121–1122. A later author on the subject is A. Kircher (1661); see 7–20 for a historic overview.
4 M.H.S., *Vom Erdpidem* (1511).
5 Luther (1883–2009) X, 1ᵉ Abteilung, 2ᵉ Hälfte, 100 (Adventspostille 1522). Luther (1883–2009) XXXXII, 364 (explication of Genesis 9: 12–16): "Sed ego nulli unquam Libro minus credidi, quam illi de meteoris, quod hoc fundamento nititur, quasi omnia ex naturalibus causis oriantur."
6 Sudhoff (1894/1958).
7 Rauscher (1911) 262, 270–271.
8 Paracelsus (1925) 395–401: *Uslegung der Erdbidem, beschehen nach usgang des Cometens in den Alpischen birgen in M.D.XXXI.*
9 Paracelsus (1925) 397.
10 Paracelsus (1925) 398, 399; cf. also 402, a draft for this pamphlet, which in the seventeenth century was published from manuscript.
11 Paracelsus (1931) 127–206. See also the introduction, vi–ix.
12 Paracelsus (1931) 167.
13 Paracelsus (1931) 274–275.
14 Rose (1980); on the *Theatrum*, Blair (1997).
15 Bodin (1597) 163–164.
16 Bodin (1597) 201–210. See also Blair (1992).
17 Bodin (1597) 177.
18 Torreblanca (1623). Feltmann (1691) refutes the actions of demons.

6

THE EMERGENCE OF A SCIENCE OF SIGNS

The comet of 1531

The deluge predicted for 1524 did not happen. In the end, the episode appears to have been an isolated event. The many strange things and wonders that people reported in the first three decades of the sixteenth century were terrifying, but were not part of a coherent worldview. Nor did Church leaders for the moment develop strategies to deal with prodigies and prognostications, either to neutralize them or to use them for their own ends. For the moment, it was not clear how these phenomena could be brought into agreement with the dominant philosophical and theological ideas.

This changed, however, with the appearance of the great comet of 1531, one of the passings of Halley's Comet. There are several reasons why this phenomenon elicited an intellectual response. In the first place, the comet was not just spectacular, it was conspicuous. Unlike most other prodigies, which were strictly local, comets are visible all over the globe and for long stretches of time. The comet of 1531 was visible from the beginning of August to the beginning of September, although most reports are from mid-August. Moreover, comets are rare. For Melanchthon, it was the first comet he ever saw. Although smaller comets had been reported in earlier years, a comet of this size was definitely uncommon.

In the second place, there existed a sizeable body of literature on comets, from classical authors like Pliny and Aristotle to the medieval scholastics. Scholars had to take position with regard to this existing literature. In the third place, right at this time, the printing press was coming to maturity. All this made that the comet of 1531 elicited a flood of writings and commentaries. Astronomers studied its location and course in the sky, mostly to come up with an astrological prognostication. Theologians and religious leaders discussed why God had sent this sign and what it meant for the present state of the world and the course of history.

In the fourth place, the political situation had made a dramatic turn. The comet appeared one year after the imperial diet of Augsburg and the writing of the so-called Augsburg confession. At Augsburg, the attempts to bridge the divide between the Church of Rome and the Lutheran reformers had foundered. Two parties emerged that would soon clash in open war. The nascent Lutheran Church organized itself and sought to strengthen its position. The Church leaders started a systematic campaign to bring the population on their side. To this end, they did not so much appeal to learned theological arguments, but rather to divine signs and wonders. Prodigies had been around for a while, but mostly as an expression of rather diffuse fears. Now they were given a distinctive meaning: telling people to embrace Lutheran doctrine and become obedient Church members. The reference to the external threats as they were manifest from (among others) natural wonders, brought their message home to a wide audience.

The debate on the comet therefore did not remain an isolated event, but was the start of a sustained publicity campaign.[1] The many treatises on the comet offered examples of how to interpret similar events, such as northern lights, of which there were some spectacular instances in the sixteenth century, blood rains, or images in the sky – armies, religious figures, dragons, or other animals, reports of which were also frequent. In the following decades, a growing body of pamphlets on all kind of prodigies appeared. As Hellmann found, the growth of meteorological pamphlets set in around 1530.[2] Most of these pamphlets were written by Lutheran clergy or other leading figures.

However, the incorporation of these prodigies into established philosophy and theology was not unproblematic. Religious leaders could not afford to let fear of the supernatural go rampant. Religious "enthusiasm" was a real threat to the established order of society. With few exceptions, the reformers did not intend to bring about a social revolution. They rather sought the support of the secular authorities, and in their turn supported the strengthening of the state. They readily appealed to prodigies of all kind, but of course at the same time denied the interpretations of their adversaries. So, a solid philosophical framework was needed that could rein in too adventurous interpretations.

Such a philosophical framework was not immediately available. Scholastic philosophy, after all, had consistently tried to explain away the prodigious meaning of strange phenomena such as blood rains or flying dragons and most humanist scholars had followed suit. The growth of a popular pamphlet literature therefore was accompanied by the development of a learned discourse on prodigies and the meaning of unusual events. Prodigies became a serious object of study for established scholars in the forefront of both the Protestant and the Catholic traditions, who worked towards creating a coherent world-view, integrating philosophical, theological as well as edifying concerns.

Here too, the comet of 1531 was a turning point. Whereas most authors on the comet restricted themselves to the phenomenon itself and its meaning, others regarded it as part of a broader issue. They came to discuss the meaning of rare events more generally. Joachim Camerarius on the Protestant side, and Fridericus Nausea on the Catholic side, used the comet as an occasion for such general treatises. In this way, they turned the study of prodigies, including earthquakes, into a branch of scholarship. They developed a general discourse that enabled Church leaders from both sides to both use and harness the popular fear by referring to such prodigies.

The Wittenberg interpretation of prodigies: Camerarius and Melanchthon

The one classical author who gave the most respectability to the belief in prodigies, was Pliny. His "Natural history" not only is a store-house of wonders, but also permeated with the belief that these strange phenomena have some meaning. In his chapter on fires in the air, he passed in review many phenomena that in the sixteenth century were considered ominous and discussed their meaning. The work also has a large section on earthquakes. In discussing the causes of earthquakes, Pliny followed the basic theory as put forward by Aristotle: earthquakes are caused by winds that are trapped inside the earth and shake the earth when they try to break out violently. But unlike Aristotle or Seneca, Pliny also explicitly stated that the danger of earthquakes was not just in the fact itself, but also in what they announced: "The city of Rome was never shaken without this being a premonition of something about to happen", a phrase that was repeated over and over again by early modern authors.[3]

The most influential scholar in the renovation of learning in Protestant Germany was Philippus Melanchthon, professor at Wittenberg. He would earn the nickname of *praeceptor Germaniae*. Though a friend and follower of Luther, he did not accept Luther's anti-naturalism and instead advocated a philosophy based on Aristotelian tenets, though accommodating for specific religious sensibilities. He was an active proponent of predictive astrology and also accepted the need to include a theory of prodigies into natural philosophy.

Melanchthon had been deeply impressed by the comet of 1531. He turned for advice to his friend, the famous Nuremberg philologist Joachim Camerarius.[4] It is probably on his instigation that Camerarius wrote a treatise on this occasion, *Norica*, a Ciceronian dialogue wherein Camerarius allegedly reports a conversation on comets in the Nuremberg circle of Johann Mylius. Melanchthon wrote a preface to the book. First, Michael Roting gives a physical discourse, wherein he demonstrates that philosophers hold different views on the nature of comets. Next, Camerarius himself argues that comets, whatever their causes, are extraordinary phenomena that must have a special significance and announce disastrous mutations. He compiles a long list of historical comets to demonstrate that all these have been followed by dire disasters.[5]

Later, Camerarius also wrote a book on various kinds of divination, but this was not published during his lifetime, probably because the Wittenberg professor

Caspar Peucer had meanwhile (in 1550) published an exhaustive work on the subject.[6] Peucer was the son-in-law and friend of Melanchthon and the two appear to have closely cooperated in writing the book. There are several parallels to Melanchthon's textbook of natural philosophy, *Initia doctrina physica*, which had been published the preceding year. Peucer's main question is what forms of divination are illicit and what forms are licit. Illicit are magic, incantations and other practices which implicitly or explicitly appeal to the devil. Three forms are licit. The first are divine oracles; the second is divination from natural causes, as medical prognosis, meteorological signs, and, the most powerful of all, astrology. Astrology was regarded by both Peucer and Melanchthon as based on natural causes and hence a part of natural philosophy.[7] The third kind of permitted predictions, with which Peucer closed his book, is *teratoscopia*, the interpretation of signs (monsters, apparitions in the sky, and the like) without a natural cause, which are sent either by God or the devil. Peucer's book has especially been discussed for its legitimation of astrology, but one might argue that teratoscopy was for Peucer and his contemporaries at least as important as the influence of the stars.

In this final chapter, Peucer expressly denied that extraordinary phenomena, like monsters and other deviations of nature, were only sports of nature, as Aristotle had considered them. The cause of deviations was not in nature, but higher.[8] God had created a perfectly harmonious order of nature, in which everything served the well-being of man, but this order had been disturbed by the Fall. In the heavens, the region above the moon, the original harmony probably still held. The heavens were free of mixed and conflicting qualities. This idea was behind much of the study of astronomy at Wittenberg.[9] But in the terrestrial realm, even if the admirable fabric of causes, by which all secondary causes are at the service of the First Cause, had not been completely disrupted, in many places it was seriously compromised. Strange phenomena occurred, generation produced deformed offspring, and the Devil had free rein.[10]

To the meteorological signs, Peucer devoted a whole chapter of his book. (In later editions, this chapter was much expanded.) As for divination, a distinction had to be made between the consideration of common effects, which led to physiological predictions (predictions from natural causes) on the one hand, and on the other the contemplation of new and unusual phenomena, which led to teratoscopical predictions (predictions from monsters and portents).[11] In meteorology, this corresponded with a distinction between meaningful meteors and meteors without meaning (*sèmantika* and *asèmanta*). The latter just follow their nature. They have physical effects, which one can therefore predict, but no special meaning. Such are the phenomena which occur daily, as rain, snow, dew, and so on. The *sèmantika*, on the other hand, occur rarely and are of a conspicuous character. Examples are mock suns and inundations, or even common meteors when they occur in an abnormal way, like snow in June. They are followed by disasters, not as a physical effect, but they announce them as a sign. The harm may regard particular persons, but more often kingdoms and republics.[12]

In general, things announce evil and horror when they transgress their natural limits, "either from natural causes, that are increased beyond their natural measure; or from God himself and the angels; or from demons, whose power is principally in the air".[13] Peucer expressly rejected those ancient philosophers who, by giving natural reasons for everything, had tried to show that God does not interfere with the world. In the study of meteorology, therefore, one had to consider not just natural, but also supernatural explanations.

The correct interpretation of miracles and portents was a question that would occupy still other Lutheran thinkers, theologians as well as philosophers, but the basic framework had been laid.[14] These principles could then be applied to the interpretation of concrete phenomena. As we will see in the next chapter, the work of Peucer had a considerable impact on the teaching at Wittenberg, and thereby throughout the Protestant world.

The Catholic interpretation of portents: Nausea

A similar debate on prodigies and different sorts of divination started at about the same time in the Roman Catholic Church. Here too, the epicentre of debate was Germany and there appeared to be considerable influence from the humanist movement.

The most important author here was Fridericus Nausea (Friedrich Grau), later Bishop of Vienna. His work too was mainly inspired by the great comet of 1531, but he had given thought to the subject already some years earlier. In 1528, when he was a canon at Mainz on the Rhine, an earthquake struck the city. At the request of Laurentius Truchsess, Dean of Mainz, Nausea thereupon wrote, "An answer on the major earthquake at Mainz of this present year 1528."[15] The first two pages (in the 1531 edition) consist of a series of short Latin poems on the subject. After some dedicatory lines and a prayer to God, these verses first briefly demonstrate that philosophers disagree about the cause of earthquakes. Next they explain where earthquakes come from, and what they portend. The main treatise then has the form of a prose commentary, still in Latin, on these *carmina*.

In the commentary, the disagreement among philosophers is documented with some learned sources, but in the end Nausea agrees to a form of the Aristotelian explanation. He continues with some examples of the havoc wrought by earthquakes and briefly touches upon some of the standard questions regarding the place, time and various types of earthquakes. Finally, there is the question what earthquakes portend. Nausea states that he is sure that earthquakes do not happen in vain, but have a hidden meaning. Referring to the text "There shall be earthquakes, in divers places" (Matthew 24:7), he states:

> So we understand from the explicit word of the Lord, that an earthquake does not happen simply according to custom, but with some meaning, and always portends to us the wrath of God, and at the same time earnestly exhorts us to improve our life and manners.[16]

The last chapter, the longest of the work, then gives a long list of earlier earthquakes and the evils they announced.

The work on the earthquake for the moment remained unpublished, but when in 1531 the great comet appeared in the sky, Nausea was among the many authors who wrote a treatise on the phenomenon.[17] It was published with a dedication to Ferdinand of Hapsburg, King of the Romans and King of Hungary. With the general interest in portents now considerably intensified, his earlier treatise on the earthquake got printed as well. It was dedicated to Joannes Faber, alias Johann Hegerlin or Johann Fabri, Bishop of Vienna. The letter of dedication is dated September 1, 1531, just about three weeks after the comet had started to become visible.[18] The earthquake treatise was well received. A German translation was included in a volume published by the mathematician Johannes Rasch in 1582, together with a translation of Beroaldo's treatise on the Bologna earthquake of 1505, and the Protestant minister Abraham Fleming gave an English version in 1580.[19]

Moreover, Nausea seized upon the occasion to publish a more comprehensive and general work on prodigies, his "Seven books of wonders", which was published at Cologne in 1532.[20] The character of this book is different from Peucer's, even if they discuss very much the same topics. The first three books are on generalities. In book one, Nausea briefly discusses why certain things are rightly called *mirabilia* (wonders). If we keep in mind the infinite power of God, nothing should amaze us. Our amazement is raised, however, when we do not know the causes of something. We are capable of learning anything, but God, in arranging the world, saw fit that we do not understand certain things, either because they are against nature or because they are uncommon. So, from the outset, Nausea makes clear that wonders are an integral part of God's plan with the world.

Book two gives an elaborate discussion of the various types of wonders that Nausea distinguishes: miracles, portents, monsters, omens, and others. Each category is illustrated by examples, both from sacred and profane history. Nausea made a point of arguing that scriptural wonders can be discussed on an equal footing with those happening at other times. Book three is important in that here, Nausea harmonizes the belief in prodigies with Christian orthodoxy. If wonders announce disaster, that is not because of their nature, but (and here he agrees with Peucer) because God uses them to warn us. However, God could have done so in any other way. Therefore, we cannot deduce any specific predictions from such wonders, unless we have the gift of prophecy. Wonders do not serve to give us knowledge about future events. They are a sign by which God shows the world his mercy or (more often) his indignation.

In book four, Nausea discusses some specific wonders. Most of these are images in the sky that had been reported recently, but he also refers to monstrous births, prodigious rains, and comets. He suggests that he has diligently collected eyewitness reports. Most reports are probably from Germany, but in one case, he refers to letters from Apulia (Italy) informing him of a local wonder, a rain of bread. Apparently, he was considered an expert on such phenomena already. As for causes, Nausea is well aware of the scholastic theories that explained such phenomena in a natural way from exhalations. In the case of religious images in the sky, he suspends judgment: God could work these by natural and supernatural means equally well. As for meaning, referring to histories of earlier, similar wonders, Nausea leaves little doubt that such events do mean something, probably something bad.

Book five then is devoted to the wonders of the "present year" 1531 and their significance. This refers especially to the great comet, but puts it in a wider framework by asking the question why the present time sees so many more wonders than earlier periods. This leads to a discussion of the impending Last Judgement and the dissolution of the world, and how we should behave under this prospect. This book was clearly intended as the conclusion of the work. The last chapter is even called its peroration. But then, Nausea's earlier treatises on comets and earthquakes were still added as books six and seven.

The treatise on earthquakes we already discussed. In the treatise on comets, Nausea rehearses the various opinions on the origin and properties of comets, but then quickly moves on to their meaning. His main message is that comets portend evil. In the case of the comet of 1531, this is true irrespective of whether it is caused naturally or supernaturally, for all natural indications point to a bad aspect: it has the colour of Mars and Saturn and stands under the sign of these two planets, according to some astronomers. Also, its course deviates towards the north, from where according to the prophet the evil comes (cf. Jeremiah 1:14, 4:6, 6:1, 51:48). So, biblical prophecy is used in a natural explanation. The chapter ends with a call for repentance and conversion.

Nausea returned to the subject of wonders still on a later occasion, in 1544. At the time, Germany suffered from an invasion of big locusts. Nausea wrote a treatise, dedicated to the Holy Roman Emperor, wherein he interpreted this as indicating war, famine, and plague, and as a divine warning to repent our sins. Compared to 1532, however, the impending end of the world had receded into the background.[21]

Nausea did not remain an isolated case. Several other Catholic authors would tackle the subject. The Jesuit, Franciscus Torreblanca published a *Daemonologia* that promised in the title that it would deal with natural, demonic, licit, and illicit magic.[22] The first part, out of four, is devoted to divination. Like Peucer, Nausea and others before him, Torreblanca tried to draw boundaries between various kinds: prophecy, natural divination, artificial divination, and divination from various kinds of prodigies. He admitted that some prodigies might have natural causes, but on the whole, he was not very interested in those.

Earthquakes he regarded as an unambiguous sign of God's anger. Torreblanca gave several examples, mostly taken from Church history (as in later ecclesiastical authors). From Nikephoros, he took the story that during the sack of Rome by Alaric, houses split into two, whereafter the two halves were joined again so that nothing remained of the damage. Such things cannot be explained naturally, he pointed out.[23] After earthquakes, he dealt with volcanoes and fissures in the earth, which he treated in a similar way.

Earthquakes recur in Torreblanca's second book, on operative magic, where he discussed the works of the devil. After having explained at some length that the devil can raise storms, he continued by saying that the devil also can cause earthquakes, either by making a vehement spirit enter the earth, or by moving the subterranean air. He emphasized that in this way, never the whole earth will move, but only a small part; according to the philosophers (Seneca) not more than 200 miles wide. However, at the Crucifixion, the whole earth trembled, clearly demonstrating that this was a divine miracle.[24]

A very ambitious programme for the interpretation of signs in nature was written by Cornelius Gemma, the son of the Louvain mathematician Reinier Gemma Frisius, and himself a professor of medicine at the same university. In 1575, he published at Antwerp a comprehensive work, *De naturae divinis character-ismis* [On the divine characteristics of nature], commonly known as *Cosmocriticus*. The occasion of writing the book was, as he explained himself, the appearance of the new star of 1572. Gemma wanted to give a theologically sound explanation of prodigies. In his view, not all things diverting from the regular and normal course of nature should be deemed prodigies, even if they are rare. This term properly belongs to the irregularities that have their origin in a special providence of God or in the depravities of fallen nature. Both normal nature and prodigies are divine, but the latter are used by God in a special way. In this way, Gemma tried to maintain the nature of the prodigious, which in other authors seemed to merge with the simply wondrous. Comets, earthquakes, and monsters according to Gemma are caused by the heavens. However, because of the Fall, they now work in an ugly way, instead of their original orderly purpose. They definitely do have meaning. The troubles in the Netherlands were preceded by an earthquake in 1554 and two comets thereafter.[25]

Gemma's work certainly did get some attention in this era so fascinated with prodigies. Examples from his book were frequently referred to by other authors. However, in the end it did not get the Church's blessing and it never became part of established authority. Its neo-Platonic elements were hard to swallow for religious orthodoxy. In spite of Gemma's pious intentions, his approach to nature appears more like Cardano's web of correspondences, than the orthodox interpretation that was establishing itself following the lines laid out by Nausea, Torreblanca, and others.

Compilations of prodigies

In the wake of the more theoretical work by Peucer, Nausea, and others, people started compiling catalogues of wonders and prodigies. Interest in prodigies was of course nothing new. Medieval chronicles abound with them. In those chronicles, however, we find the prodigious events interspersed into the records of deaths, births, successions, wars, and other historical events. In the era of the Reformation, catalogues were compiled that listed only prodigies. The standard was set by the chronicle of prodigies by Julius Obsequens, a late classical author, as edited and expanded in 1552 by Conrad Lycostenes (or Wolffhart). Lycosthenes followed up with his own *Prodigiorum ac ostentorum chronicon* [Chronicle of prodigies and monsters, 1557]. Similar works were published by Marcus Fritsche in 1555, Job Fincel in 1556, and Caspar Goltwurm in 1557.[26] These works were often reprinted and saw many imitators. A somewhat later example is *Zeitbüchlein* [Booklet of the times], published in 1580 at Jena by Valentinus Rudolphus, a Protestant cleric and schoolmaster, which related what between 1500 and 1580 had happened "as for wars, famines, signs in the heavens and on earth...".[27]

Some authors mainly compiled classical data from authors and historians, but others focussed on prodigies of their own times. The choice of events in these works appears often rather haphazard. The genre drew its popularity, no doubt, from the combination of sensational content, partly concerning distant lands, and the clues offered for understanding the course of history and the coming of the Day of Judgement. Moreover, such books had an outspoken edifying purpose. A treatise on recent signs and portents from 1613 closes with a list of historical examples of persons who had taken no heed of such signs and ended badly, and another list of persons who reacted exemplary with prayer and humility.[28]

Catalogues like the above-mentioned listed all kinds of unusual events: comets, earthquakes, monsters, strange meteors, and so on. But early on, some authors specialized in one particular type of prodigy. Best known of these are the catalogues of comets. Camerarius already included such a list in *Norica*. Years later, when Camerarius was professor at Leipzig, he returned to the subject in a disputation, wherein his stated aim was, by systematically recording how earlier comets had appeared and what had followed them, to be able to determine the meaning of future apparitions. In the seventeenth century, many comparable lists would be compiled.[29]

Earthquakes as well had their own collectors and compilers, even before the sixteenth century. Scholastic treatises normally ended with a list of notable earthquakes. The mid-fifteenth-century humanist scholar Gianozza Manetti dressed a catalogue wherein he listed 210 earthquakes since the creation. Nausea included a similar (though much shorter) list in his treatise on the Mainz earthquake of 1528. Later authors continued this practice. Many meteorological treatises or disputations, as well as pamphlets or sermons on particular earthquakes, include a list of historical earthquakes. For instance, the minister Johannes Moltherus included a catalogue with his sermon on the earthquake of 8 September 1601.[30] Nicolaus Höpffner published in 1691 "the shaken and trembling Meissen and Thuringia", on the occasion of an earthquake which had struck central Europe in November of the year before. Of 78 paragraphs, 41 are devoted to the history of earlier earthquakes.[31] An academic disputation defended in 1684 at Bremen has as its aim the physical explanation of earthquakes, but about half of the work consists of a catalogue of noteworthy earthquakes of the past, the quakes mentioned in the Bible taking pride of place.[32]

These sixteenth and seventeenth-century compilations have a different character than the earlier ones. The scholastic or Renaissance lists were historical or philosophical exercises. The catalogues that were compiled from the late sixteenth century onward, however, did not aim to understand earthquakes in themselves, but rather to document the warnings God had sent to mankind. They were edifying works rather than works of scholarship, although not infrequently they were composed by people of learning. The authors might be priests or ministers, but also professors or amateurs. Obviously, these were all works of compilation. They served to underline that people had to repent and prepare for the Last Judgement. In many cases, like the contemporary catalogues of comets, the catalogues aimed to demonstrate that earthquakes always had announced something evil.

From the late sixteenth century onward such compilations were also published as separate works. One of the first was occasioned by the earthquake of Vienna of 1590: *Chronica ... von den Zornzeichen Gottes den Erdbidemen* [Chronicle of the signs of God's anger, the earthquakes] by Valentinus Fabricius, published at Erfurt in 1592. The full title is telling enough:

> Chronicle wherein recorded a true and thorough description of the terrible miracle works and gruesome signs of God's anger, the earthquakes, how they always mean some future event. Useful to read for all pious Christians. With a short admonition what opinion everybody shall have of earthquakes. Compiled in chronological order....

Fabricius explained that he had written the book so that nobody would think that the stories on the earthquake were just figments of the imagination. Unlike some later earthquake chronicles, the work is not very long. The chronicle itself is thirteen pages, the preceding admonition eight.

Fabricius started with some brief remarks on portents in general. He emphasized that they indicate God's anger and warn us of his punishments, as well as being signs of the impending Last Judgement. Some of them are punishments themselves. Because such signs are wonders that happen without and often against nature, they should teach us that God is not bound to secondary causes. Signs like images in the sky, as they are so often seen in the author's time, happen without any natural cause. Other signs do have natural causes, "but these one should not scrutinize or investigate, but one must first of all have regard of the final cause, wherefore and why they happen". All wondrous signs have one efficient cause in common, to wit, God. Scholars have assigned natural causes to earthquakes, especially the influence of the planets, "but these are not sufficient nor complete, for earthquakes do not happen without purpose [*ohne gefehr*], but they always signify future disaster...". The list of earthquakes appears somewhat hastily compiled, as it is not overtly accurate. Moreover, the list is highly selective. Earthquakes that do not imply future punishment are lacking, including all earthquakes from the New Testament.[33]

Ten years later, the lawyer Johann Michael Beuther at Strasbourg published a *Compendium terraemotum or Erdbidems Verzeichniss*, a catalogue of earthquakes from the birth of Christ to his own time. Beuther explained on his title page that he wrote his book on the occasion of the central European earthquake of 8/18 September 1601, "as a necessary consideration for many, what to think of earthquakes", and based on the records of "both Church- and other reliable historians". Just as happened frequently in the case of catalogues of comets, Beuther not just listed the phenomena themselves, but also went into their significance, in pointing out the events that followed or provoked them. So, the earthquake in Philippi in 50, mentioned in Acts 16:26, that liberated Paul and Silas from their dungeon, showed "that God almighty works earthquakes, when two righteous Christians and preachers are persecuted". When in 70 AD, the emperor Nero was deposed, an earthquake showed "that the earth too no longer would suffer such a tyrant". Another earthquake accompanied the rise of Mohammed.

Beuther also paid ample attention to the earthquakes of his own time. The meaning of the earthquake of 1580 in the Netherlands was not hard to decipher: "there is still no peace in the Netherlands". The disaster of Yvorne in 1584 was followed by war between Bern and Savoy, the earthquake of Vienna of 1590 by an Ottoman invasion and persecutions of the Protestants in the Austrian lands. As Beuther explained, all these examples made it not hard to understand what we should feel of the earthquake of 1601: "Since every time war, famine, persecution, and other major disasters have followed, we should expect these now as well." Unless, that is, we pray to God for mercy and mend our ways. Beuther closed his book with some prayers to say at the time of an earthquake.[34]

In the seventeenth century, such earthquake chronicles still mushroomed. An anonymous (and rather chaotic) chronicle published at Hamburg and running till 1692, had the title:

> Chronicle of calamities, of many fearful and horrible earthquakes which since the beginning of the world (...) have happened, to everybody's utmost wonder and amazement, up to this time, in which the great miraculous finger of God has made itself felt especially in this epoch full of wonders.[35]

They appeared not just in the German lands. The year 1691 saw the posthumous publication of a catalogue by the Italian nobleman Marcello Bonito, recording all earthquakes in history from the creation to his own time. Bonito took up this project after he had barely survived an earthquake at Naples in 1688, but the result was not a mere occasional work, but a thorough study of over eight hundred pages.[36] The Italian Franciscan friar Filippo di Secinara published in 1752 a treatise "of all the earthquakes that happened, and are known, in the world, with the unhappy and unfortunate events that were predicted by such earthquakes", a work of some 150 pages.[37] In England, a comparable list of earthquakes was published in 1750 "by a gentleman of the University of Cambridge".[38] Another list, from the earthquake at the Crucifixion to the year 1750, was added to the French translation of the book by the German philosopher Johann Gottlob Krüger on the ancient history of the earth.[39]

Such catalogues were works of piety rather than of natural understanding. Interestingly, they are normally not much concerned with the distinctions that are made by learned authors, like the distinction between natural and supernatural events. They regard earthquakes as portents, but they include miracle stories as well. Of course, the authors knew the difference. Abbati in 1703 gives a list of miraculous earthquakes, "produced by the almighty hand of God the Creator, independently of any natural cause", but admits that there are others that are the mere effects of natural causes.[40] However, what counts is that earthquakes announce God's message. The treatises by Peucer, Nausea and others should be seen as, not very successful, attempts to harness the general awe of portents into a theologically acceptable form, rather than as real guides for the perplexed.

Portents and miracles in Church history

To wonders like earthquakes was given one general meaning: repent your sins and mend your life. In practice that also meant: be a good Church member. In earthquakes, God showed his displeasure with heretics or with people who despised true religion or combated the true Church. Theologians appealed to such signs to carry home the message that people should remain within the flock and not break lines.

Especially the Christian historians and chroniclers of late antiquity, who themselves were constantly combating pagans and dissenters among their own, offered a storehouse of relevant stories. These were eagerly taken up by the historians and compilers of catalogues of the sixteenth and seventeenth centuries. One of the main works of the Counter Reformation was the *Annales ecclesiastici* by the cardinal Cesare Baronio. It is a huge work in twelve volumes, published between 1588 and 1607, wherein the complete history of the Church up to 1198 is described from a Counter-Reformation perspective. (As such, it was a reply to the Magdeburg Centuries by Flacius Illyricus, who did the same from a Protestant point of view.) Baronio's work would remain an authoritative source for later historians and compilers. It is a work of impressive scholarship, based on the study of a wide array of ancient sources, but of course it is also decidedly partisan.

The book emphasizes God's guidance throughout history. God constantly and actively interferes in the course of events to protect his Church. Apart from Popes, heresies, doctrinal differences and the like, Baronio therefore paid ample attention to prodigies and miracles. We find several stories that figure prominently in many later earthquake catalogues. So, Baronio related how during the great earthquake of 446 of Constantinople, everybody fled into the fields. The earthquake did not cease. Suddenly, a child by some divine power was lifted into the air and heard the angels singing. Having landed back on earth, he taught the bystanders the song he had heard: "Holy God, Holy Powerful God, Holy Immortal God, have mercy on us". The bishop told the people to sing the hymn, and the earthquake ceased at once. The child thereupon died. The song, the Trisagion, would remain one of the main hymns of Catholic liturgy. Baronio's main source for this event was the historian Nikephoros, but he felt the story to be important enough to corroborate it with a host of other sources.[41]

Rather than just stating in a general way that earthquakes are signs of God's anger, Baronio does not hesitate to attribute specific causes to them: God is angry about heresy. So, it was no coincidence that the Trisagion was revealed exactly in 446: God provided it because Eutyches was secretly preparing his new heresy. On the earthquakes of 553 in Constantinople, Alexandria, and other places, Baronio comments:

> Who can deny, if he can judge the facts from their outcome, that these were presages of the evils, that followed upon the synod held that year? For then the Churches were rocked with a heavy concussion, which offered to the cities of east and west a pitiable spectacle when they collided upon each other....[42]

In this way, a pattern becomes visible, whereby especially the enemies of the Church are struck by earthquakes, whereas the Church offers protection against them.

Another earthquake discussed at some length is the earthquake in Antioch of 525. It is clearly ascribed to human sin. According to Baronio, the city had "prostituted" itself with heresy, until finally, it received from God's hand the punishments for its sins and drank the chalice of God's anger.[43] There is also a long story of the earthquakes in Antioch in 528. In their anxiety, the inhabitants implored divine protection. One of the inhabitants, moved by a divine oracle, told his fellow-citizens to inscribe above the doors of their houses: "Christus nobiscum: state" (Christ is with us: stand!). And indeed, the earthquakes ceased abruptly. The city was thereupon renamed Theopolis, "city of God".[44] These were not considered to be mere stories of long ago. They actually provided guidance to people who needed to come to terms with earthquakes in their own times. The phrase: "Christus nobiscum, state" was written by the inhabitants of Lima in Peru above the doors of their houses as a protection against earthquakes, for which this city was notorious.[45] Several later earthquake treatises include prayers that are borrowed from these stories.

Other Catholic authors too used earthquakes to defend the supremacy of the Church. Torreblanca related that the first earthquake in history was the earthquake under King Uzia, reported in the Old Testament (Amos 1:1 and Zacharia 14:5).

> For even if we should admit that there had been the crime of idolatry, as well as other crimes, throughout the world, never had God made his anger manifest by an earthquake, unless the freedom of the Church had been violated by the Kings.[46]

The Antwerp Jesuit Jacobus Tirinus argued, in his commentary to Amos 1:1 and 2 Chronicles 26:22, that this earthquake must have taken place in his 27th year, when the King, having grown arrogant because of his victories, usurped the priestly office and brought sacrifices in the temple.[47] Earthquakes, in other words, showed how much God was on the side of His Church in its struggle with secular powers.

One might expect that Protestant authors were less prone to refer to this kind of miracle stories. After all, the Protestant reformers rejected Catholic miracle-mongering as impious and idolatrous. However, when it came to the interpretation of signs and wonders, Protestants too, including theologians, would thankfully draw from the rich store of ancient miracle-stories and interpret the various signs in that light.

Sigismund Suevus (Siegmund Schwabe) was a prominent Lutheran churchman in the second half of the sixteenth century. His "Mirror of human life" gained wide popularity as an edifying book. It consists of a number of treatises, one of them on earthquakes. This treatise, probably originally a sermon and first published in 1581, was written on the occasion of two earthquakes in Leipzig and Vienna (1578, resp. 1581). As common, among much else Suevus gives here a list of histories and examples of earthquakes. The first is the above-mentioned earthquake under King Uzia, which Suevus dates at 787 BC. Like Tirinus, he explains that many Church fathers state that this earthquake happened when Uzia sacrificed in the temple. This shows "how those who usurp Church offices by force, commit a sin that is so horrible and grave, that the earth abhors, shakes, and is moved".[48]

Suevus too regarded this as indicating a general pattern. Rome saw an earthquake in AD 242, after a rebellion against the emperor Gordian, showing that God hates rebellion. When under the emperor Julian the Apostate, the Christians were persecuted, God destroyed the pagan temples at Antioch and Delphi with thunderstorms and earthquakes. In 1226 occurred an earthquake in Lombardy when there was a conspiracy against the emperor Frederick and people tried to impede his crusade. He also refers to the earthquake of Constantinople in 438 (which Baronio put at 446)[49] and the singing of the Trisagion, and to an earthquake in 344 AD in Pontus wherein only the church and the house of the bishop remained standing, as an example that we can keep faith in God.

Other Protestant authors too included these stories. The Lutheran pastor Nuber gave a long list of such miracles in his sermons on the earthquake of 1655. Among others, he included the stories of the earthquakes of Constantinople 446 (Nuber writes 444), with the miracle of the child, and Antioch 528, with the formula written above the doors.[50] Beuther as well includes the miracle of the child (dated to 438) and among the prayers he included to say during an earthquake is again the Trisagion. Clearly, Protestants were not always averse to miracle stories.

An earthquake that was told by various Protestant commentators is the earthquake that marked the reign of the Roman Emperor Julian the Apostate. In 363 AD, the emperor had decided to rebuild the Jewish temple at Jerusalem, to spite the Christians. Christian chroniclers relate how when the work neared completion, a major earthquake destroyed the building. Obviously, this was a divine vindication of the Christian cause. In case anybody might miss the point, the historians adorned the story with a number of other miracles: phantoms, fire from the sky, etc. The story was retold many times, both in edifying texts and in academic contexts. It is the subject of a German *Meisterlied* from 1626. At a disputation at the University of Wittenberg, the case was listed under the supernatural earthquakes. There is still an elaborate version in the Dutch physico-theology by Bernard Nieuwentijt from 1715, a very popular work in the first half of the eighteenth century. (However, it is lacking in the English translation that appeared three years later.)[51]

Whereas Luther had insisted to see the whole world as an expression of God's will, later authors felt compelled to write systematic treatises wherein they distinguished between meaningful and meaningless phenomena and between natural phenomena, miracles, prodigies, and so on. Thereby they accomplished several important goals. They made clear that extraordinary events are not fortuitous, but should be regarded as signs sent by God. This allowed the respective Churches to use these signs as a means to convey to the common people their message of sin, repentance, piety and redemption. Moreover, they did so in a definite orthodox framework. Earthquakes and other events proclaimed God's will, but only along lines approved by religious orthodoxy. They showed God's displeasure with immoral behaviour, the rise of heresy, or attacks on the rights of the Church.

By rejecting the possibility that these signs meant anything more specific and could be used to predict the future, theologians prohibited unwelcome interpretations by laymen. By making a clear distinction between natural and non-natural phenomena, they put a barrier against the radical anti-naturalism of Luther and Paracelsus, which might likewise have upset the social order. The interpretation of signs remained firmly in the Church's hand. In practice, this meant that the reading of events as *exempla* was discouraged. Any strange phenomenon is teaching us God's general message. It should not be read as telling us something about specific worldly affairs. The Churches never denied God's active providence. Indeed, their interpretation of history was permeated with it. But at the same time, they did promote a view of the world as governed by the natural properties of things.

Miracle stories were traditionally the domain of theologians, historians, and preachers. But with the establishment of the Churches of the Reformation and Counter-Reformation, the interpretation of such signs became an affair of philosophers as well. The philosophical understanding of the world had to answer to the new confessional demands.

Notes

1 As also argued by Barnes (2016) 141, who however focuses on astrology.
2 Hellmann (1921) 13.
3 Plinius (1979) 331 (Nat.Hist. II, 200).

4 Rauscher (1911) 260, 266–267; Kusukawa (1995) 125–126, 134–135, 167–170. On Camerarius, see Woitkowitz (2003) 32–46; see esp. 39 for the circle of Mylius. On his interest in astrology: Brosseder (2004) 105–106.
5 Camerarius (1532). On *Norica*, see Ludwig (2003) 18–29; Freyberger (1987) 38–43; Céard (1977) 170–174.
6 Peucer (1553). The first edition is from 1550; the book saw in all nine editions between 1550 and 1607. Some later editions are considerably expanded. Ludwig (2005) 20–34. See also Müller-Jahnke (1992); Weichenhan (1995); Céard (1977) 178–186; Barnes (2016) 138–139; Roebel (2004) 220–240.
7 Methuen (1996) 396–397; Brosseder (2004) 247.
8 Peucer (1553) 322–323. Melanchthon would largely follow Peucer's argument on monsters, see Melanchthon (1550) 143v–146.
9 Methuen (1996) 398–400.
10 Peucer (1553) 324–325. Cf. also the chapter on meteorology in later editions: Peucer (1602) 568–569.
11 Peucer, (1553) 239. Cf. Peucer (1602) 573.
12 Peucer, (1553) 246–246v.
13 Peucer, (1553) 249v.
14 Ehmer (1988) 187–188 discusses the systematic treatment of miracles by the theologian Jakob Heerbrand.
15 Nausea (1531), introduction. N.B.: This introduction was only added in 1531.
16 Nausea (1532) fol. 74 v.
17 The treatise is discussed in Rauscher (1911) 273–276.
18 Nausea stated in the introduction to the comet book that the comet "primum (ni fallor) circiter XIX. cal. Septembris hoc nimirum anno post-Christum natum MDXXXI admodum portentose apparere coepit…".
19 Nausea (1531). Rasch (1582); A. Fleming (1580).
20 Nausea (1532). Ewinkel (1995) 35–36.
21 Nausea (1544).
22 Torreblanca (1623). The imprimatur etc. of the book is from 1612 to 1613, the dedications are dated 1618.
23 Torreblanca (1623) 26–28.
24 Torreblanca (1623) 224–225.
25 Céard (2008) 71, 73 and *passim*. Gemma (1575) II, 23, 180.
26 Laureys (2005) 210–212; Schwegler (2002) 35–41; Schilling (1964); Céard (1977) 161–162, 186–191; Boaistuau (2010).
27 Rudolphus (1580).
28 Leuchterus (1613) 46–48.
29 Mosley (2013).
30 Moltherus (1601). The catalogue is on pp. 37–52, the actual sermon on pp. 6–36.
31 Höpffner (1691).
32 Disputation Bremen (1684 May 3). Other examples: Ragor (1578) 36–61; V.A.D.L.C. (1580) section 10–50 (no page numbers; this list starts chronologically, but ends chaotically); Ziegler (1674) 7–8; Grimaldi (1703) 39–72. See also Van de Wetering (1982) 422 note 19, on eighteenth-century Puritan sermons.
33 Fabricius (1592) (no page numbers).
34 Beuther (1602) (no page numbers).
35 *Unglücks-Chronica* (1692). Cf. Eckstorm (1620) for still another example.
36 Bonito (1691).
37 Secinara (1752). Some earthquake chronicles are discussed in J.G. Taylor (1975) 229–241, but mostly for their theoretical content.
38 *Chronological and historical account* (1750).
39 [Krüger] (1752) 243–328.
40 Abbati (1703) 5.

41 Baronio (1738–1746) VII, 595–598. See also Dagron (1981) 96.
42 Baronio (1738–1746) X, 341 (year 553, ccl).
43 Baronio (1738–1746), IX 343 (year 525, xiii). See further xiv–xviii on the same earthquake.
44 Baronio (1738–1746), IX 379 (year 528, annotation xxi–xxii.) For a modern interpretation of the events, see Meier (2004) 229–231. The story of the inscription had already been told by Maggio (1575) 65v.
45 Feltmann (1691) 8.
46 Torreblanca (1623) 27, referring to Hieronymus' commentary on Isaiah 7.
47 Tirinus (1632) I, 635; II, 753. See also Magnati (1688) 5–6.
48 Suevus (1587) 274v–289v, quote on 277. See also Hoffmann (1927) 112–113.
49 There is actually confusion about the true date of this earthquake, see Croke (1981).
50 Nuber (1655) 36.
51 Brunner and Wachinger (1986–2009) VIII, 329. Disputation Wittenberg (1608 Febr 27) th. 6. Nieuwentijt (1715) 522–524.

7
PRODIGIES IN REFORMATION SCHOLARSHIP

The Reformation was a program to reform the Church, to reinstate piety, and to create a god-fearing society. The Catholic Counter-Reformation, somewhat later and in a different way, had very much the same goals. The program went further than reforming Church doctrine in a strict sense. Piety and purified religion had to determine everybody's life. In the works of nature too, people had to recognize God. The science of signs as formulated by Peucer, Nausea, and others had a practical purpose. Its aim was to set the limits wherein people were taught piety and obedience to the Church. As such, it was a knowledge destined to be divulged into the world, in sermons, edifying treatises, histories, and other works. At the same time, as a new form of orthodoxy, it needed the continuing support of the intellectual elite, both philosophical and theological. The interpretation of strange events was defended in university textbooks, academic disputations, and other works of learning. Hence, we can indeed speak of a "science of signs".

The science of signs was on the one hand popular and edifying, on the other scholarly and educational. The two sides were closely related, but not identical. Each field presented its own demands and limitations, and each field had its own means of communication. In the academic world, language and protocols were to a large degree standardized throughout Europe. On the other hand, the means of edifying the population at large had many regional and confessional differences. The two fields therefore have to be treated separately. Here, I will deal with the science of signs as an academic and scholarly discipline. In a following chapter, I will discuss how these views were communicated to the population at large. For reasons that will become clear, Catholics and Protestants have to be treated separately.

The transformation of learning in the Protestant world was carried out at several levels. High priority was given to the training of a new intellectual elite.

The sixteenth century saw the founding of many new universities and the reformation of old ones. Teaching henceforward followed the principles of the new reformed faith. The contents of these programmes can be studied from the textbooks in use as well as from student disputations. Disputations were set exercises, wherein a student defended a number of theses, typically on a standard question from the curriculum. The theses themselves had normally been written by the presiding professor. In these works, the principles as set out by Melanchthon, Peucer, and others had to be implemented. Authors had to find a balance between on the one hand religion, which demanded that one recognized God's hand in nature, and on the other hand philosophy with its legacy of naturalism.[1]

The field where these questions most naturally were discussed was obviously meteorology. It will be remembered that in the medieval tradition, meteorology explained wondrous phenomena in a purely naturalistic way, without referring to direct divine intervention. This changed in the sixteenth century. Both Protestant and Catholic philosophers seized upon Aristotelian meteorology as a major support for their view of the world, but both reformulated it in a way that recognized the role of the supernatural in terrestrial events. The sixteenth and seventeenth centuries saw both a strengthening of the position of the field of meteorology in the academic curriculum, especially in a confessional context, and a reinterpretation of this field. In this way, the Reformation and Counter-Reformation had an important impact on philosophical and scientific thinking.[2]

The Wittenberg tradition in meteorology in the sixteenth century

In the wake of the Lutheran reformation, the University of Wittenberg became the leading university in the Lutheran world. The professor Philippus Melanchthon set out a programme for the reform of academic teaching that was imitated at many other universities. Initially, Wittenberg had abandoned the teaching of Aristotelian physics, mainly under the influence of Luther and his aversion of Aristotle's naturalism. By the mid-century, Melanchthon again returned to the teaching of Aristotle for the basic curriculum. Still, it was an Aristotelianism in a specific Reformed mould. The aim of propaedeutic teaching was not just to give an introduction to basic Aristotelian theories, but also to get an idea of God's activity in the world.[3]

In spite of Luther's misgivings about meteorology, Lutheran philosophers, exactly because they wanted to show God's hand in nature, and because they attached so much importance to phenomena like comets, prodigious rains, and so on, found they could not ignore the field. A central question was to what extent the various phenomena were natural, and to what extent they were direct interventions of God. Actually, meteorology became much more prominent at Wittenberg than it had been at medieval universities.

For some time, physics at Wittenberg was mainly taught from the second book of Pliny's Natural History.[4] In 1543, Professor Jakob Milich, a close collaborator of Melanchthon, published a new edition of this work, with extensive commentaries which made it fit as a textbook. Pliny's second book is devoted to the heavens and the universe, meteorological phenomena, and the earth. This inevitably gave meteorology a prominent position in the curriculum. The choice for Pliny may well have been inspired by the fact that, unlike Aristotle or for instance Seneca, Pliny did *not* want to explain everything from natural causes. His works are full of stories of apparitions, presages, and other miraculous occurrences.

Milich certainly was aware that his time was full of signs. He refers to fiery meteors he has seen, and also to an earthquake in his hometown of Freiburg in Breisgau on 1 November 1509.[5] He elaborately discussed all kinds of fiery meteors, such as comets and northern lights (*chasmata*), but also less common ones as the sound of armies in the air, the raining of blood, flesh, or iron, and the like. On the other hand, he already appears to move back to a more philosophical position. He left no doubt that comets and some other phenomena announced future events, but this did not mean they had no natural cause. He rather saw an analogy to the predictions from the stars, by astrology. Such indications of the future showed that the world was governed by a Mind.[6]

With regard to miracles, Milich was rather cautious. Since everything depended on a Mind, meteors were not just formed by chance. Physically (or materially), they signified some mutation of the air. Experience showed that they were signs as well, but their meaning was more obscure than the astrological meaning of the stars. He, however, preferred to consider just their physical goals.[7] In most cases, he abstained from giving a cause at all, but just referred to the various authorities. So, on the wonder of two colliding mountains, related by Pliny, he commented that, although it was not possible to attribute firm reasons to such portents ("for they are produced by nature as portents and prodigies"), Albertus Magnus had still investigated their natural explanations. "If someone desires those, they can find them in Albertus."[8]

As stated before, Pliny's book II also has an extensive discussion of earthquakes. In his commentary, Milich passes in review most of the standard points from scholasticism. He emphasized the astrological cause of earthquakes (or of the exhalations), as could be expected at Wittenberg. Moreover, he did refer to the final cause as well. Earthquakes signified some future event. This was not just asserted by Pliny, but also stated in the Scriptures, in particular 2 Samuel 22.[9] Moreover, true to the spirit of Pliny, he paid much attention to all kinds of wondrous events that happened in the earth. His views seem to be not just inspired by the strict biblicism of Luther, but also to have something in common with the worldview of Cardano.

Starting in the middle of the sixteenth century, several textbooks were published specifically devoted to meteorology. In 1545, the poem on meteorology by the Italian neo-Latin poet Giovanni Gioviano Pontano, which already in 1524 had seen a publication at Wittenberg, was published with a commentary by the Wittenberg professor Veit Amerbach for use by the students. The first real textbook, a short "compendium from Aristotle, Pliny and Pontano" was written by

Johann Lonzer (Lonicerus) and published in 1548. He was followed by Michael Stanhuf in 1554, whose book saw a new, much enlarged edition in 1562. Probably the most important, certainly the most comprehensive, meteorological textbook was written by Johannes Garcaeus, first published in 1568. A second edition from 1584 was still more comprehensive.[10]

Garcaeus was in his time a prominent scholar, close to Melanchthon, especially active in the field of astronomy and astrology. He is best known for a vast compilation of horoscopes. His meteorology goes well beyond what is required of an academic textbook. It is edifying as well as scholarly and might well have been used by ministers preparing a sermon. He explicitly stated that he had written the book "for the service of students, to glorify the prophetic and apostolic writings, which cannot lack physical knowledge".[11] In his general principles, he largely followed Peucer. He made the same distinction between *sèmantika* and *asèmanta* and denied that all meteors had natural causes. Moreover, unlike the medieval authors, Garcaeus does not appear to feel a real difference between contemporary meteors and biblical miracles, as they are discussed on an equal footing. Among the meteors he includes are not just the earthquakes at the Crucifixion and the Resurrection, but also the fiery column which preceded the Israelites in the desert, the Deluge, and the star of Bethlehem which announced the birth of Christ, which are all absent in the medieval meteorologies.[12]

In the chapter on earthquakes, Garcaeus followed Aristotle's theory as regards their causes, effects, and other characteristics, and rehearsed the familiar scholastic points, but he also included a section on their final causes. Their final cause was the announcement of some major future disaster. The Bible teaches that by means of these phenomena God shows His anger, announces punishment, and calls for repentance. Because of this view of earthquakes as divine instruments, Garcaeus also lists "notable examples of people who have been swallowed by the earth or have fallen into gaps and precipices in the earth"; and: "memorable examples of earthquakes in the Sacred Pages". He ends with a catalogue of major earthquakes collected from histories, 165 in all, with the wars, plagues, or other disasters that followed them. "In Germany we have seen similar examples these days, whereupon have followed fatal mutations in the Empire."[13]

Wittenberg set the tune for most of the universities in Protestant Germany, including the teaching of meteorology. Textbooks on meteorology soon were published at other places as well. Jodocus Willich published a short outline of Aristotle's meteorology at Frankfurt on the Oder in 1549.[14] Still another meteorological textbook was published by Markus Fritsche at Nuremberg in 1555, significantly together with a catalogue of prodigies. After theological disputes had led to the expulsion of many of the professors from Wittenberg (Peucer even was incarcerated), the new professor Johannes Hagius in 1581 republished Fritsche's book at Wittenberg, claiming to have it reworked, but as far as could be verified, he only changed the order of the chapters.[15]

At the University of Leipzig, the main rival of Wittenberg, two other textbooks were published, curiously enough in the same year (1587). The one was by the theologian, Henricus Decimator, the other by the learned physician, Wolfgang Meurer. Meurer's book was published posthumously by the author's son and republished in 1592 and 1606. The title page explicates that the book contains the subject of Meurer's earlier lectures at Leipzig. Meurer emphasised that meteors show God's greatness and works. They have a function in cleaning the air, but also in announcing future events. He too followed Peucer's lead in making a distinction between *sèmantika* and *asèmanta*.

In the chapter on earthquakes, Meurer discusses thirteen *quaestiones*, with answers, followed by thirteen *explicationes*.[16] He shows himself an erudite scholar, touching on many different aspects. In several points, the influence of Georgius Agricola, whose friend he had been, is apparent, especially in his emphasis on the role of the subterranean fire. One of the questions concerned, here again, the final cause: earthquakes signified God's anger. Although they were no direct *cause* of future harm, they certainly announced it, as was shown by the histories. In the explication, Meurer stated that there could be no doubt about this signification of earthquakes, as this is stated explicitly in the Bible. Like Milich, he referred to 2 Samuel 22. Earthquakes particularly teach us that nothing is so firm that it cannot be overturned.[17]

All of these textbooks pay much attention to prodigies and final causes: unusual meteors have special meaning, unlike common meteors that just serve their function in the order of nature. On the other hand, they stand in the Aristotelian tradition of explaining the phenomena from natural causes. This leads to a somewhat uneasy tension between Luther's anti-naturalism and the philosophical tradition in the explanation of prodigies. Most authors agreed with tradition that many strange phenomena, even if they should be regarded as signs, could be explained by natural causes. On the other hand, they would insist that at least some of these phenomena were beyond the powers of nature and should be regarded as miracles, although they did not always agree on what these phenomena were. The biblical miracles in any case were direct acts of God. Since earthquakes figure prominently among these, there was a strong motive to regard other earthquakes as miraculous as well.[18]

Lutheran meteorology in the seventeenth century

In the seventeenth century, the tradition of meteorological textbooks petered out in the Lutheran world. A compendium that appeared in Frankfurt in 1619 was a mere summary of Aristotle's work.[19] However, meteorology remained a topic at universities. Earthquakes and other phenomena were dealt with in several academic disputations. In general, disputations were set exercises that just rehearsed received knowledge, but sometimes, professors used them to take a public stance on questions that were not in the standard curriculum. Consequently, they are rather diverse. Meteorology was a frequent topic of disputations at Lutheran universities, attesting to its continuing appeal, but it is dealt with in a variety of ways. Some disputations simply follow the medieval tradition and focus on the physical causes. Others, in the Wittenberg tradition, focus on prodigies and the prodigious nature of the phenomena, whereas still others, though philosophical, focus on biblical themes. (In practice, of course, the boundaries are blurred.)

Among the "physical" explanations, there is one defended at Rostock in 1596 under the professor Nicolas Andreas Gran. In the dedication, the student, a Swedish nobleman, explicitly praised meteorology for showing that many things, contrary to vulgar opinion, were natural, and thereby proving Seneca's word that philosophy took away superstition. In the disputation itself, Gran occasionally rejected vulgar opinion. He argued that comets and their effects were purely natural. They could even be predicted from the position of the planets. Fiery meteors seemed prodigious, but they were so not by chance or necessity, but from nature. On the whole, Gran was only interested in natural explanations. Only in the case of rainbows did he explain the theological significance.[20]

Rudolf Goclenus was a prominent Lutheran philosopher, teaching at the university at Marburg in Hessen. In 1599, he presided over a disputation on fiery meteors. He distinguished between divine and natural causes, but he discussed only the latter. The things worked naturally by a faculty ordained in them by God. The *ignis lambens*, a light settling itself on a person's head, was by many people regarded as completely prodigious without natural cause, but Goclenus insisted that looking more closely, the phenomenon did appear to have a material and efficient natural cause. He did discuss the final cause of the meteors, but was silent on them being signs announcing evil. Instead, their goal according to him was the purging of the air from smoke. He was more tolerant of astrological meaning and gave the significance of shooting stars according to pseudo-Ptolemy's *Centiloquium*.[21]

A professor who had a somewhat isolated position in early Lutheranism was Nicolaus Taurellus. He was professor of philosophy and medicine at Altorf, in the territory of the imperial city of Nuremberg. In later years too, Altorf would often steer an independent course with regard to the other major Lutheran universities. Taurellus' independence did not only show in his theology, but in his meteorology as well. In 1602, he presided over a disputation on earthquakes. Herein, he gave a detailed critique of the Aristotelian theory. He preferred the theory of the pre-Socratic philosopher Anaxagoras (as reported by Aristotle) and felt that rather than in wind or air, the cause of earthquakes was in the earth itself. The earth was full of cavities and when overhanging rocks fell down, they shook the earth. The disputation is nearly exclusively physical. Only at the very end, Taurellus included some remarks on the goal of earthquakes. Although he conceded that God and nature do nothing in vain, he denied that earthquakes were of any use to man. "We feel that it happens by a necessity of matter and of acting causes, rather than to serve anybody." He refused to believe that earthquakes announced anything. However, if the earth were ever to perish naturally, one could tell beforehand from the many earthquakes.[22]

His namesake Martinus Taurellus presided over a disputation on winds at Wittenberg in 1606. It has one section on earthquakes. Herein, he said that they caused fear only so long as we are not aware that they are God's work. No better remedy to the superstitious fear of earthquakes, than to learn the causes. "For sure, fear always draws and leads us to a desire of knowledge and understanding." This is an interesting twist to a well-known Stoic mantra, as the causes he refers to are obviously no physical causes. Still, it is remarkable that the fear of punishment appears to have a small place.[23]

More in the tradition of sixteenth century textbooks is a series of nine disputations on meteorology defended under Tobias Tandler at Wittenberg. They were published together in 1609. Tandler mostly discusses the physical causes, but does not ignore religious elements. In the third disputation, on fiery meteors, he asks whether the devil can work such meteors and answers affirmatively. In the eighth disputation, he discusses whether the rainbow did exist before the Deluge, a much debated question among Lutherans. As to earthquakes, their causes are partly natural, partly supernatural (*extra naturae*). That God sometimes causes earthquakes is proven from the Bible.[24] Another disputation on earthquakes from 1608 emphasized that some earthquakes are supernatural rather than natural; an example of the former was the earthquake under Julian the Apostate.[25]

A central question in many discussions on earthquakes was whether the earthquake at Christ's Passion had indeed been world-wide, as was traditionally believed both by Catholics and Protestants. The (supposedly) universality of the earthquake was considered a proof of its miraculous nature. A disputation on meteors defended at Wittenberg in 1606 had as one of the corollaries "whether there has been another universal earthquake, apart from the one during Christ's Passion?" The question was answered in the negative, therewith underlining the exceptional character of this earthquake.[26] Tandler the following year agreed that there was no natural power that could move the whole earth. The earthquake at the Passion was universal, but it happened by God's infinite power that is bound to no laws.[27] In all, opinions at Lutheran universities were definitely more diverse than a mere study of the Wittenberg textbooks would suggest, but the interpretation of prodigies, the natural or supernatural character of phenomena, and the use of the Bible in philosophy were important themes.

Other Protestant denominations

Academic textbooks on meteorology appeared especially in the Lutheran world. In other Protestant countries as well, theologians and philosophers were concerned with the interpretation of meteorological phenomena, but they did not explain them in systematic scholarly treatises. A number of other sources however give us some access to their ideas.

In England there appeared some works on meteorology in the vernacular. One of these, *A goodly gallerye*, was written by William Fulke, a prominent Puritan divine and controversialist. The book came out in 1563 and was reprinted numerous times, the last time in 1670. Fulke's book follows the traditional Aristotelian framework. He was reluctant to admit divine miracles and used meteorological theories to combat popular superstitions. For instance, prodigious rains were explained by means of the standard scholastic explanations. Fulke also rehearsed a story that around 1547, people had seen the devil flying over the Thames, a case apparently still well-known at this time. He supposed this must have been some fiery meteor: "Thus do ignorant men iudge of these thynges that they knowe not."[28]

He also included a lengthy chapter "Of wonderfull apparitions", wherein he discussed the images that many people saw in the sky. He did not deny that these were signs, "in dede miraculus" sent by God as exhortations to mend our ways; "but not yet so, that they want a natural cause". Fulke tried to explain the apparitions as an optical phenomenon, a kind of mirage. Because the mirrored images could be deformed or combined with other images, the resulting images could be quite phantastical. However, he emphasized that even so, such phenomena were wrought by God's Providence: "thoughe I haue enterprysed to declare these by naturall reason, yet beleuing that not so much as on sparrow falleth to the grounde, without Gods prouidence, I doe also acknowledge Gods prouidence bryngeth these to passe…". Fulke justified his rejection of the divine character of apparitions in the sky with the consideration: "if they be well weyghed and considered, it is not harde to finde, that they differ much from such miracles, as ar recorded in the scripture, and admitted of divines".[29]

In the end, Fulke certainly does not want to remove God from nature. Whereas the common phenomena such as rain and dew are dealt with only briefly, he devotes ample space to phenomena of a more spectacular or prodigious character: fiery meteors, thunder and lightning, earthquakes, and apparitions. He gives many examples of their wondrous effects and definitely sees them as signs of God. He concludes the part on earthquakes by reminding his reader: "But what lande, can be sure? if it be the Lordes will, by this woorke of his to shake it?"[30]

Other English works were less well balanced. Thomas Hill published in 1571 *A contemplation of mysteries*. This book has the form of a meteorological treatise, but is to a large extent an enumeration of prodigies. Meteors are instruments used by God to punish sinners. Many do not have a natural cause but are plainly miraculous. As to earthquakes, Hill does include the standard topics, but also their final cause: "the signification verie sad & heavie, of matters and haps to come, as of battels, landfloodes, mutation of Emperies, the dearth of victuals, &c. For the Earthquakes always pronounce great calamities".[31]

A vernacular work that stands out in the period is the epic by the French Huguenot Guillaume de Salluste, lord of Bartas, *La sepmaine* (the seven days), on the creation. The work saw many reeditions and translations well into the seventeenth century. Bartas left no doubt about the meaning of prodigies. Prodigious rains, for instance, he described as infringements upon the course of nature and as Divine warnings. (One should notice that this holds not true for the raining of frogs, which Bartas regarded as a perfectly natural.)

> *Dieu, le grand Dieu du ciel, s'egaye quelquefois*
> *A rompre haut et bas de nature les loix*
> *Voulant que les effects, à nature contraires,*
> *Soyent les avant-coureurs des futures miseres.*[32]
>> ["Sometimes, it is the pleasure of the great God of heaven to disrupt nature's laws above and below, when he wants the effects, contrary to nature, to announce future miseries."]

He discussed comets, blood rains, and so on. On earthquakes, he remarked first that the element of earth always remained in its place and was a support for humanity. Still, it was true that sometimes God, in his anger over the sins of man, made a small part of the earth move, "with help of the northern winds, that, as if imprisoned in its hollow intestines, grumble and rave".[33]

Biblical earthquakes remained mostly a topic for theologians. Reformed theologians held much the same line as their Lutheran colleagues. Benedictus Aretius, a theologian from Bern in Switzerland, wrote a commentary on the Acts of the Apostles. Originally published in 1579, it saw several re-editions. The earthquake of Acts 16:26 was a sign of God's answering the prayers of the apostles, although Aretius did not omit that in other cases, earthquakes signify God's anger. His main point in discussing this earthquake was that it was not natural, but supernatural and extraordinary. The earthquake was not conform to what philosophers had found, and experience taught, to be the common characteristics of earthquakes. In normal earthquakes, high buildings are more shaken than low ones, as seen in the effects of a small earthquake at Bern in 1569, April 5. Very heavy earthquakes can shake the foundations as well, but in that case, whole buildings will collapse. According to Acts 16, however, only the foundations were shaken, whereas the building itself remained un-damaged. (As an aside, the Jesuit Tirinus argued that the earthquake had been felt throughout the city, as in some Greek manuscripts, it was found that the magistrate was so terrified that they set the apostles free, seeing "that the ele-ments and God were fighting [*certare*] for them".[34]) Aretius concluded that God showed His miracles to save his followers and to frighten the unbelievers.[35]

At Zürich, the professor of theology Johann Hottinger in 1652 published a volume of historico-theological treatises, including one "on the judgments of the Jews and the Arabs on earthquakes", based on an earlier disputation. It is an erudite treatise with many references to ancient authors. In large part, it relies on an earlier writing in Hebrew that a certain rabbi Azarias had written on the occasion of the 1570 earthquake of Ferrara.[36] Questions discussed among others are whether Amos 1:1 is about a real earthquake, or is meant metaphor-ically (Hottinger prefers a literal sense); whether this earthquake was the same as the one that happened under King Uzia, when the King wanted to usurp the office of a priest; and as the one mentioned by Jesaja 6:4 (no). A long section is then devoted to the question whether earthquakes were regarded as natural or supernatural events. The Jews traditionally understand them as acts of God, but Hottinger again largely follows rabbi Azarias, who had defended a middle course. Some earthquakes are natural, some supernatural, and even when they are natural, God can use them for a supernatural end. There is a long list of the various sins that earthquakes punish, and their theological goals.[37]

Philosophers too emphasized the divine character of earthquakes. Claude Aubéry, a Swiss physician and philosopher who also wrote on theological questions and was at the time Reformed, held an academic oration on the occasion of the disaster of Yvorne (1584), in the territory of Bern. He tried to keep a balance between natural and supernatural causes. Natural causes in the end depended upon God, who made use of them but could work without them as well. Consequently, in his oration, he replied on the one hand to people who attributed everything to natural causes but ignored God's hand in the events, or defended false causes which ran counter to theology. On the other hand, he answered people who attributed everything to God and denied natural causes, "as if first philosophy or theology turns the other sciences completely upside down, instead of constituting and stabilizing them".[38] So, even if wind or spiritus was the nearest cause of an earthquake, they were so only because God wanted them to. In discussing the natural causes, he followed the familiar scholastic elements. However, in the end, this should all be attributed to God.

The prominent Reformed philosopher Bartholomäus Keckermann held a disputation on earthquakes at the University of Heidelberg, occasioned by the great earthquake of Unterwalden in September 1601. It was printed several times. The dedication is dated 20 January 1602. For a disputation, the work is rather long. Keckermann started by explaining that it was not unpleasing to God if we tried to understand such unusual phenomena, since this would give us greater insight into the effects and thereby in God's works.[39] The first part is then devoted to a description of the earthquake according to the Aristotelian classification. In the second part, Keckermann discusses the efficient causes. He distinguishes between natural earthquakes, supernatural earthquakes (like those at the Crucifixion or the Resurrection), and those whereby natural and supernatural causes are both present; that is, where God gives the natural causes an extraordinary force and intensity that goes beyond their nature. Some historical earthquakes were of this kind and according to Keckermann, the 1601 earthquake too belonged to this class of "extraordinary and wondrous earthquakes".[40] This especially because of its extension, its unusual location, and its unusual velocity. In order to make his point, Keckermann makes a long digression on the natural causes of earthquakes, relying on the Aristotelian theory (but including subterranean fire). He admits that the astrological constellations and the weather were favourable for generating the 1601 earthquake, but in the end he concludes that the subterranean spiritus cannot be held responsible for such a large concussion without extraordinary assistance from God.

The final, third, part discusses the effects of the earthquake and what it predicts. As the causes are both natural and supernatural, the 1601 earthquake both has natural effects and is a warning ordained by God. The natural effects are principally disease, inundations, and infertility of the soil. However, since God does nothing in vain, the earthquake also has a special meaning that does not follow from its nature. "It happens so that by these wonderful and hidden works of nature, because of some ordination and council of God, the human minds are for one part awakened and forewarned, for another terrified because of fear of future events."[41] There follows a long discussion of what earthquakes in general, and the one of 1601 in particular, might mean. Keckermann discusses biblical earthquakes, but also miraculous earthquakes from history. He ends with a discussion of earthquakes as an announcement of the Last Judgment. Pious people regard the last earthquake as such. The treatise ends with a discussion of ten other points regarding earthquakes, both of a special and a general character. These need not detain us here.

Notes

1 Kusukawa (1995); Methuen (1996); Vermij (2010).
2 On seventeenth-century meteorology, see Martin (2011); Meinel (1987) 58–66 (discussing mostly Jesuit authors).
3 Kusukawa (1995).
4 Kusukawa (1995) 136–137; French (1986) 261–262.
5 Milich in Pliny (1543) 98, 173. An expanded edition was published in 1563 and republished in 1573.
6 Milich in Pliny (1543) 91. Cf Melanchthon (1550) 19v.
7 Milich, in Pliny (1543) 86, see also 99–99v.
8 Milich, in Pliny (1543) 176v. Cf Pliny, Nat. Hist. II 199.
9 Milich, in Pliny (1543) 172.
10 Pontano (1524); (1545); Lonzer (1548) (on Lonzer, see Dannenfeldt (1972)); Stanhuf (1554); (1578); Garcaeus (1568); (1584).
11 Garcaeus (1584), dedication.
12 Cf. Ducos (1998) 70–71 (but see also 403).
13 Garcaeus (1584) 392v. The chapter on earthquakes is chapter 41, pp. 385v–392v.
14 Willich (1549).
15 Fritsche (1555)a; (1555)b; (1581).
16 Meurer (1606) 379–392, 392–417.
17 Meurer (1606) 384, 401–402.
18 Vermij (2010).
19 Rotmann (1619).
20 Disputation Rostock (1596) dedication and theses 8, 13–15, 12, 6. On Gran, see Guliano (2019), on his meteorology see p. 119.
21 Disputation Marburg (1599), pp. 7, 20, 60–61.
22 Disputation Altorf (1602) th. 54.
23 Disputation Wittenberg (1606 Jan. 8) th. 5.
24 Tandler (1607) disputatio III problema 3; disp. VIII prob. 2; disp. IX th. 8–9.
25 Disputation Wittenberg (1608 Febr 27).
26 Disputation Wittenberg (1606 Oct. 18) coroll. 3.
27 Tandler (1607) disputatio IX th. 7.
28 Fulke (1979) 93–95.

29 Fulke (1979) 38, 84.
30 Fulke (1979) 88.
31 Hill (1571) 72v.
32 Salluste (1977) line 779–782. See also lines 739–750 and 514–526 (frogs), 767–794 (prodigies). Cf. Céard (2013) 42–43.
33 Salluste (1977) line 415–416.
34 Tirinus (1632) III, 261.
35 Aretius (1579) 31r, 77v–78r.
36 Apparently, this refers to *Me'or Enayim* by Azariah de' Rossi, on whom see Weinberg (1991).
37 Hottinger (1652). Cf. disputation Zürich (1651).
38 Aubéry (1585) 4.
39 Keckermann (1607) 151, theor. 8.
40 Keckermann (1607) 162, theor. 23.
41 Keckermann (1607) 191, theor. 67.

8

MIRACLES AND METEOROLOGY AMONG CATHOLIC SCHOLARS

The Protestants had come round to Aristotle out of necessity, but could feel free to criticize him. Catholic philosophers were more constrained, as the Council of Trent had officially sanctioned Aristotelianism as the only permissible form of philosophy. On the other hand, whereas the Protestants denounced the miracle-mongering of the Catholic tradition, with its many saints, relics, statues, etc., the Catholics stood by their tradition. If anything, the Counter-Reformation only strengthened the support. The referral to miracles was promoted by the Church leadership and supported by learned scholarship. So, Catholic thinkers were expected to show loyalty both to Aristotelian naturalism and to the rampant belief in miracles.

Whereas the Lutheran tradition of meteorology was strongest in the sixteenth century, the Catholic counterpart took off only in the seventeenth. However, there did exist a tradition of meteorological textbooks in some of the vernacular languages. As a kind of Catholic counterpart to the Huguenot Du Bartas, we can mention another French author, Pierre de La Primaudaye. In 1593–1594, he published a large overview of the visible and invisible things in the world, *Academie francoise*. Although not denying natural causes, he emphasized that in everything, we have to look for the first cause. Comets and other prodigies are heralds of the impending Judgement, calling on people to convert. Thunderstorms could be raised, by God's permission, by evil spirits. As to earthquakes, the earth cannot have any natural motion. Any movement of its parts is forced. La Primaudaye rehearsed the various natural causes that had been put forward by the learned, especially focussing on the analogy with gunpowder and siege mines. He concluded that

> if we desire a short and sure way to find the real cause, we have to refer to the anger and judgements of God, since whatever the causes that can be

found by the most learned, the Eternal One always shows himself strong and terrible by them, as he has disposed them and they depend all from him alone.

Nature in itself is not able to do anything. It is God who draws from his treasures the earthquakes, "either by a direct command, or by his ministers whom He ordains so, or by some virtue infused into the things, which can be put to effect at will, to announce his judgments onto men".[1]

Jesuit meteorology

As in the Protestant world, the creation of a Roman Catholic philosophy that would account for God's activity was primarily the task of academic teaching. Especially the Jesuits were active in the formulation of meteorological theory. The most authoritative series of philosophy textbooks in the Catholic world was produced by the Jesuits of the college at Coimbra in Portugal. Their work consisted of a series of commentaries on all the major works of Aristotle. Herein, they followed the medieval tradition of Albertus Magnus, Thomas Aquinas, and others, but brought up to date and in a way better suited to the didactics of the modern period.

The part on meteorology does extensively discuss prodigies, but most of this is relegated to an introductory part. How is it that some meteors do portend something? First, the commentators made a distinction between phenomena that have natural causes, such as comets, and phenomena that are the work of divine virtue, angels, or demons, such as the images of armies in the sky. The latter type always indicates something new and remarkable, but the former too may announce important events. Comets presage big events partly in a natural way, because they are a sign of exhalations that cause plague and other evils (the medieval scholastic theory), partly because this meaning has been assigned by God. They have natural causes, but God has arranged the natural order in such a way, that they are produced at the proper time. So, even though the author follows the medieval physical explanation, he emphasizes the role of God as the first cause who directs the secondary causes.[2]

In the discussion of the phenomena themselves, the commentators make the same distinction between natural and supernatural phenomena, but concrete examples of the latter are mostly limited to Biblical miracles – as in the case of the Lutherans, the inclusion of Biblical examples is a marked deviation from the philosophical tradition. So, the Deluge could not have been effected by the force of nature alone, but was caused "by the power and on the command of God, who applied natural agents, strengthening them beyond their ordinary power". The Star of Bethlehem was not an ordinary comet, but "some new and extraordinary meteor, not natural, but angelical, or by divine virtue conflated from the subcelestial matter, and impregnated [*conspersum*] with the fulgor of a big light". The Manna that served the Israelites in the desert for nourishment is not related to the common form, but of an angelic nature.[3]

This at the very least left open the possibility that some contemporary events too were miraculous. Indeed, in the commentary on Aristotle's book on the heavens, the author came to speak of the nova of 1572, which he had observed himself. He discussed at length the various explanations that had been put forward and rejected them all. In the end, he was left with the opinion "that this new star has been created by God not by a physical, but by a supernatural generation; but what it portends, and for what reason God has shown this unusual spectacle to the world, remains unknown". He defended himself against the objection that he had recourse to a miracle, claiming that it had been shown that such phenomena, happening outside the ordinary course of nature by divine virtue only, had earlier been seen in the sky.[4]

The eleventh treatise of the commentary on meteors deals with earthquakes. This chapter mainly rehearses the traditional scholastic arguments. The various types, effects, etc. are illustrated by many examples, both from ancient and more recent history. The commentator defends the traditional Aristotelian doctrine, but with some qualifications. He points out that earthquakes are not just caused by exhalations, but also by subterranean fires and by wind. (Subterranean fires are discussed in the next, twelfth treatise. This discussion shows the influence of Agricola.) In some cases, they happen because of the collapse of caves. As a whole, the treatise has a purely physical character, but here too, Biblical earthquakes stand apart. The author points out that only the earthquake at Christ's Crucifixion shook the whole earth.[5]

The commentary of the Conimbricenses, as the Jesuits from Coimbra are called, remained leading for later discussions of Jesuit philosophers. They stuck to the scholastic, and therefore naturalistic interpretations of phenomena, although emphasizing God as the first cause of everything. On the other hand, they also included in their philosophy discussions on phenomena from the Bible, as the Star of Bethlehem, the Manna, and the earthquake at the Passion, which were clearly regarded as miraculous. So, their attempt to include theological principles in a work on natural philosophy results in a somewhat hybrid product. On the one hand, the discourse tends to demonstrate (maybe unintentionally) that biblical miracles are a category of their own; on the other, in some cases, the miraculous can spill over into the everyday world.

The one contemporary phenomenon that they univocally regarded as miraculous were images of armies and other visions in the sky. The Jesuit professor of theology Christobal de Castro, who taught at Salamanca, discussed such images in his commentary on the twelve lesser prophets, published in 1615. The book included an introduction on the various kinds of divination – natural, artificial, and prophetic. Among other things, he discussed meteorological phenomena, notably comets, but also images in the sky. According to him, some unnamed astrologers argued that these images had to be natural, as real miracles were rare and happened mostly among people who knew the true religion, whereas these images were frequently seen among barbaric nations and had the same effect as comets. (In other words, why would God perform a miracle for those heretics if a comet would do the same job?) De Castro admitted that these arguments did not sound implausible and that some images might be explained by natural causes, but he insisted that most of them were completely miraculous.[6]

The Jesuits developed a strong tradition in meteorology. The order elaborated a philosophy that took regard of the Bible while upholding a literal interpretation. The best known example is their refutation of the Copernican system, but this was not the only such instance. Within the order, there was a constant interaction between mathematicians, philosophers, and exegetes, with the exegetes taking the central role.[7] In line with the example of their confraters of Coimbra, they kept more to the side of scholastic tradition and were hesitant to accept miracles in their philosophy – with the exception of biblical miracles.

We have a number of disputations on meteorology from Jesuit professors at Catholic German universities. In 1602 at Ingolstadt, Johannes Dannemeyr presided over some "philosophical assertions on meteorological impressions". Three years later at the same university, Conradus Reihing presided over "physical assertions on the heavens and the meteorological impressions". At Freiburg in Breisgau in 1627, Georgius Reininger had "a compendium of peripatetic doctrines collected from the meteorological books" defended. It includes an approbation by the dean of the theological faculty: "This meteorological dissertation clearly displays the admirable power of God, that from such base matter of vapours and *halitus* works such wonders." In the general remarks with which Reiniger started his disputation, he among other things discussed the final cause of meteors.[8]

These works generally follow the standard explanation, that is, they follow the scholastic Aristotelian explanations, but do include some new insights, as for example on the explanation of the Milky Way. In this way, the authors kept their work more or less up to date with contemporary developments in the sciences. Religious concerns, however, are guiding. Biblical miracles stand apart. As to contemporary wonders, the authors take some pains to steer a middle course between on the one hand attributing everything to nature, and too much credulity on the other. So, in some cases at least they leave open the possibility that prodigies have non-natural causes. Apart from the case of images in the air, this is the case for comets and for hailstones in human shapes. In most cases, however, they prefer to remain uncommitted.

The author who really started the Catholic tradition of meteorological textbooks was Libert Froidmont, or Fromond, a Jesuit from the southern Netherlands. His interest in meteorological phenomena had shown itself already in 1618, when he published a voluminous treatise on the comet of that year. In 1627, while serving as primary professor of philosophy at the *Falco*, one of four colleges of the University of Louvain, he published at Antwerp his *Meteorologicorum libri sex*.[9] In his preface, he self-consciously announced the newness of his undertaking. Others had written on meteorology, but more briefly, or they had just followed the text of Aristotle.

Fromond's book follows a traditional pattern. It distinguishes the various meteors in fiery meteors (with a separate book on comets), winds, aqueous meteors, and optical phenomena. All phenomena are discussed following an Aristotelian framework, dealing with material, formal, efficient and final causes, in that order. Like the earlier Protestant textbooks, Fromond skipped the phenomena discussed in Aristotle's fourth book. In all, Fromond's book is a learned work wherein he tried to cope with more recent insights as good as possible, sometimes refuting them, sometimes integrating them within his Aristotelian framework. Moreover, the whole work was made subservient to a pious interpretation of the world.

There is ample discussion of phenomena of a prodigious nature. Apart from the book on comets, there are separate chapters on prodigious rains and on rains of frogs. Fromond remains undecided whether such rains are really "prodigious". They are called so, sometimes because they really are prodigies, sometimes because the people just believe so. On rains of frogs, he prefers to be credulous rather than ridiculous. In the last book, on optical phenomena, there is a chapter *de spectris & phasmatibus*, on images in the sky. On these images (mostly of battles), he was more explicit. Although sometimes clouds could take on a certain shape by accident, there were too many reports that could not be ascribed to mere chance, but "where any sane man sees the direction and hand of some Intelligence". These images warn us of impending battles and defeats. Fromond briefly refuted the view of "atheists", who tried to come up with a natural explanation by explaining these images as mirrored images of things that are really happening on earth: "To what point does impiety alienate the human mind from reason!"[10] On the other hand, Fromond rejected any supernatural meaning of phenomena like firedragons, will-o'-the-wisps, and the like.

His discussion of earthquakes again follows a familiar pattern, but Fromond elaborates in many instances. He follows the standard themes, discussing them at length with many references to the learned literature, but also with arguments from observations and common experience. Most interesting is what he says on their efficient causes. In the introductory chapters of his book, Fromond subscribed to Aristotle's theory of exhalations as produced under the influence of the sun. He discussed and rejected at length the theory put forward by Agricola and Lydiat, that these exhalations were produced by the subterranean fire.[11] However, this seems to be something of a disclaimer. At a closer look, the *halitus* as Fromond describes it has little resemblance to its Aristotelian counterpart any more.

In the chapter on earthquakes proper, Fromond states that the main efficient cause of earthquakes consists in a sulphurous and igneous spirit, which escapes from the bowels of the earth. The major part of this spirit finds its origin in the subterranean fire. He clearly thinks of it as some explosive substance. In a later, separate chapter, he even makes the by now familiar comparison between earthquakes and siege mines and other explosive devices. He admits that art here surpasses nature, as the subterranean spirits are more subtle than gunpowder and therefore less expansive. However, he expressly denies that his theory about this spirit is contrary to Aristotle and even claims that Aristotle himself had mentioned the subterranean fire in this respect. Moreover, "the winds and the exhalations produced by the sun help the earthquake, and maybe they produce them also sometimes on their own, even if rarely".[12] The moon also contributes: when it is full, it multiplies the spirits by its heat. Fromond is rather sceptical about the influence of Saturn or the occult influences of the heavens in general, however. Quite unlike the Medieval Latin meteorologies, Fromond also includes a chapter on the final cause of earthquakes, which according to him is the fear of God. By earthquakes, God wants to remind us of the great earthquake that will happen at the end of times. Untypical for a philosophical work, this chapter has partly the form of a prayer.[13]

Other Jesuit authors followed in Fromond's footsteps. One of the most authoritative textbooks on meteorology in the Catholic world was published by Niccolo Cabeo. It is a voluminous work in four volumes, published in Rome in 1646 and several times reprinted. It has the traditional form of a commentary on the Meteorology of Aristotle, wherein the subject matter is discussed in the form of a series of *quaestiones*. The book's ambitions are well expressed in the title: "in which not only meteorology is explained, both from the words of the ancients and mainly from experiments of singular things, but also a near-universal experimental philosophy is presented".[14]

Cabeo used the Aristotelian meteorological framework to offer a general explanation of terrestrial phenomena, based largely on experience and general scientific insights. He critically discussed the various standard explanations and although he generally refused to dismiss them, he often came up with slightly different answers. Moreover, some of the questions that made up his text discussed topics that fell outside the standard framework. In the chapter on earthquakes, there are questions on such traditional topics as the efficient cause of earthquakes, on place and time, duration, and preceding signs, but also one on siege mines.

Cabeo was clearly influenced by the science of his time. He emphatically rejected the theory that the earth was a living being. Earthquakes were caused by a *halitus* that expanded when it caught fire. However, one did not need to suppose an elaborate system of subterraneous caves. Rather, the source of earthquakes was in veins of sulphurous, bituminous, or nitric material. In spreading, earthquakes could only follow those veins. Since most of the veins pointed either east–west or north–south, earthquakes also followed that direction. (Fromond had mentioned the same idea, referring to Keckermann, but had refused to believe it.) Cabeo proved himself sceptical about the signs which allegedly preceded earthquakes, as the experience was lacking and the causal connection in most cases not clear.

Cabeo also discussed a number of prodigious phenomena. Images in the sky according to him were definitely supernatural. Although they consisted of natural things, like clouds and vapours, these could not have obtained their shape in a natural way. "So we must conclude, that they are formed by God by means of the ministry of intelligences."[15] Like Fromond, he refuted the view that such images are created by clouds working as mirrors as the opinion of atheists. The chapter on earthquakes, however, is exclusively on physical explanations and does not refer to biblical or moral issues.

There were several other Catholic meteorologies in the seventeenth century, but they do not all need a detailed discussion. The French Jesuit Jean Baptiste du Hamel wrote a book "on meteors and fossils" that was published in Paris in 1660. Meteorology was also included in works on physics generally, as the four-volume work by Honoré Fabri. One cannot say that these are only slavishly following their predecessor. On the contrary, Fabri criticizes Fromond on several points. But they certainly wage a debate within pre-set limits. The best-known Jesuit work on the earth is by the polymath Athanasius Kircher. However, this is not a textbook. Because of its different character, we will postpone discussion to a later chapter.[16]

A late and somewhat unusual example of the Jesuit tradition in meteorology is the *Meteorologia philosophico-politica* by the Austrian Franz Reinzer.[17] The first edition is from 1698 and the book was reprinted in 1709 and 1712. The work had its origin in a disputation by one of Reinzer's students and most likely he destined it for the students of Jesuits schools. The work is richly illustrated and suited to the taste of a noble audience. The work is divided in a number of dissertations, which again are divided into *quaestiones*. Each *quaestio* consists of a number of physical conclusions, followed by a political conclusion, accompanied by a large-size emblematic illustration.

As a work on physics, the book is not very innovative. It is mostly a compilation of older theories, rather traditional Aristotelian. On the Milky Way, Reinzer argues that it consists of a great number of very small stars, but he does so on the basis of a long scholastic argument. Only at the very end, he mentions in passing that one can see so by using a telescope. The section on the rainbow is equally scholastic and omits nearly all reference to mathematics.[18]

The political conclusions draw a moral lesson from the earlier descriptions. They are clearly addressed to rulers and princes, in particular to the Holy Roman Emperor Jozef I, to whom the book is dedicated. The work explicitly connects scientific knowledge with piety. For example, the section on volcanoes draws a comparison between these mountains and powerful men: magnates must not be irritated. The section on caverns and subterranean rivers has as political conclusion: virtue cannot remain hidden. The illustration shows a river that disappears underground, with the inscription: It will emerge again. Of the dissertation on earthquakes, the first question has the political conclusion: laws are untied when they are tightened too much; the illustration shows a fire burning under a mountain on which a fortress is built. The last (fourth) question of this dissertation has the moral: piety is the fulcrum of kingdoms.

Apart from the moral lessons which are drawn by way of allegory, natural phenomena may have a significance of their own. Reinzer pays a lot of attention to meteors traditionally regarded as meaningful, like comets, hailstones that resemble animals or other things, fiery meteors, and the rainbow. There are long discussions of their significance. Comets announce in a natural way draught, earthquakes, plague, and famine, and probably in an indirect way war, rebellion and the death of kings, all in agreement with traditional scholastic theory. But besides that, they announce the death of kings, war, and other disasters because that is how God wants it. This is clear both from the Bible and from historical examples. The fact that comets are natural does not mean that they do not mean anything. All this is stated in the physical, not in the political part.

Reinzer also has a discussion of images in the sky. He admits that some of them may originate in a natural way, by reflection or otherwise, like the images that are produced by a magic latern. This explanation however does not hold for images at great height. These consist of vapours and exhalations that are not formed by nature, but by intelligences in God's service. God uses these images to announce imminent disasters.[19]

Non-Jesuit works

Interestingly, textbooks written by non-Jesuit Catholic philosophers appear to pay more attention to prodigies and wonders. The Franciscan priest Franciscus Resta published a textbook on meteorology at Rome in 1643. Resta too tried to keep to Aristotle, but the biblical and theological elements are prominent. He devotes a long section, six chapters, to the subterranean fire, a topic that was at the centre of interest after the great eruption of Vesuvius in 1631. After a long discussion wherein he weighed all possible explanations, Resta concludes that the fire we see erupt from volcanoes is really the fire of Hell as it has existed in the centre of the earth since the creation.[20] The idea that Hell was probably at the centre of the earth was common in medieval theology (it was stated by Thomas Aquinas as a probable opinion) and the identification of volcanoes with the mouths of Hell must have been familiar in popular belief or preaching, but this is the first time this idea emerges in a work of philosophy. Torreblanca had described the fires of Etna as a divine warning, but stuck to the traditional physical explanation.[21]

In his chapter on earthquakes, Resta discussed whether earthquakes are natural or supernatural. Cautiously, he admits that in many cases they are natural, but without excluding the possibility of a supernatural cause. The chapter is mainly a compilation of many sources, following the traditional scholastic classification. He includes a section on the meaning of earthquakes – earthquakes as signs. Resta does not explicitly state what should be considered licit and illicit signs or predictions, but in practice he clearly draws the lines. He is critical of the traditional idea that earthquakes portend disaster. As natural events, they can mean only their own effects. Astrological and other predictions are also dismissed. However, "to these inane significations must be opposed those which we read in the prophets and the stories of the saints".[22] Earthquakes are often indications of God's anger. They are often punishments for sins, especially to avenge attacks on the independence of the Church. Sometimes, the trembling of the earth indicates God's majesty and presence. Earthquakes accompanied Christ's miracles; they demonstrate the presence of a saint; they may signify that God listens; sometimes the earth trembled when a great Saint died; and so on. Resta's book claims to be a work of philosophy, but it refers as much to the Bible, the Church fathers, and to works of theology and hagiography, as to philosophical authors.

The Benedictine Benedictus Mazzotta taught philosophy and theology at the University of Bologna. He published a textbook on philosophy, "natural, astrological, and mineral", in 1653. Among the final causes of meteors, the prediction of future events figures prominently. Portents, or prodigious meteors, sometimes have natural causes, sometimes not. Comets, although they by God's will signify future events (not always bad), have natural causes. The images that people sometimes see in the air do not have such causes and are wrought by angels or devils, under God's free will [*beneplacita*]. Like Resta, Mazzotta too feels that the subterranean fire is the fire of Hell and purgatory. It has no natural character, but has been placed in the earth by God to punish sinners. It does not burn and cannot be extinguished. It is from this peculiar character of the fire, that the most likely explanation derives how there can still be snow on the tops of volcanoes.[23]

As to earthquakes, Mazzotta feels that they are caused by minerals evaporating under the influence of the subterranean fire, whereas the external cold impedes these vapours to escape. As a corroboration, he adduces some chemical experiments he has performed. He discusses at some length the question whether there can be natural causes of earthquakes. It cannot be denied that some earthquakes are caused by God, angels, or demons, but others are caused naturally. He then rehearses the standard list of questions. It deserves notice that when it comes to the effects of earthquakes, he refers to Cardano's opinion that earthquakes announce war, plague, and tyranny. There is a separate chapter on mountains and islands, wherein Mazzotta, among other things, explains that although most mountains have been created by God to adorn his creation, some have been made later, by earthquakes, wind, humans' effort, angels or demons, or the subterranean fire.[24]

A crucial question in these debates was where to put the borderline between the natural and the supernatural. Nobody wanted to do away with the natural altogether. Everyday phenomena were governed by mere natural forces. The supernatural did exist, but was an infraction into the ordinary world and should fit the orthodox theological framework.

What everybody agreed on were the miracles in the Bible. These served as the model for miracles in general and were included as such in works on natural philosophy. Contemporary miracles made clear that one still lived in the same world, governed by the same principles, as Christ and the Apostles. Protestant and especially Lutheran authors tended to interpret certain natural phenomena, most notably earthquakes, in this light. These were on a par with the miracles of the Old Testament. Catholics, in particular Jesuits, tended to be more cautious in this respect and preferred to deal with natural phenomena as distinct from those mentioned in the Bible. They definitely accepted miracles, but tended to see these in stories about saints, relics and so on. One might argue that Protestants, because they rejected saints and relics, for contemporary miracles had to have recourse to the world of nature, whereas Catholics, who had the supernatural still very much around them, were more free to study nature on its own terms.

Neither saw natural philosophy as just the study of nature. Philosophy had to account for the whole of reality, of which the world of the Bible was an integral part.

The question remains to what extent this confessionalized philosophy was actually of any use when it came to the explanation of actual earthquakes. The various theories were not just an academic training in logical argument, but actually meant as instruments to help ministers and theologians to put these phenomena in an orthodox framework and explain them to their flocks. The interpretation of actual earthquakes is the subject of the next chapters.

Notes

1 La Primaudaye (1593–1594) III, 423, 425; see also 355 and 340–341 (comets and thunderstorms).
2 *Commentarii collegii Conimbricensis* (1600) (Meteor.) 9–10, 31.
3 *Commentarii collegii Conimbricensis* (1600) (Meteor.) 94, 34, 77.
4 *Commentarii collegii Conimbricensis* (1600) (Cael.) 72.
5 *Commentarii collegii Conimbricensis* (1600) (Meteor.) 130.
6 de Castro (1615) 17–18.
7 Remmert (2007) 157–173.
8 Disputation Ingolstad (1602 Sept 25); disputation Ingolstad (1605 March 18); disputation Freiburg in Breisgau (1627 April 21).
9 Fromond (1627); see also Fromond (1656). On the book, see Vanpaemel (2014); Meinel (1987) 62–64.
10 Fromond (1627) 416.
11 Fromond (1627) 24–33.
12 Fromond (1627) 200, see also 198, 209. The chapter on earthquakes is on pp. 196–220.
13 Fromond (1627) 216–217.
14 Cabeo (1646). On Cabeo, see Martin (2011) 106–124.
15 Cabeo (1646) 223.
16 du Hamel (1660). Fabri (169–1671).
17 Reinzer (1709). On this book Meinel (1987).
18 Reinzer (1709) 73.
19 Reinzer (1709) 99–102.
20 Resta (1643) 134–161.
21 Torreblanca (1623) 28–29. See also Vermij (1998).
22 Resta (1643) 504. The chapter on earthquakes is on pp. 471–512.
23 Mazzotta (1653) 45–46.
24 Mazzotta (1653) 147–151.

9

REACTIONS TO EARTHQUAKES IN THE SIXTEENTH CENTURY

The emergence of a discourse

So far, the discussion of earthquakes had remained mostly a theoretical exercise. The interest in prodigies was mostly focussed on comets, apparitions in the sky, northern lights, monsters, and other events that people encountered frequently. Earthquakes to most people were unfamiliar phenomena. Pamphlets brought news of earthquakes in foreign lands, but for the moment, there was little occasion to pay special attention to them.

This situation changed in the last decades of the sixteenth century, as several large earthquakes occurred in Europe which hit major cities. People asked for explanations and interpretations and many authors complied. These authors had to decide in what terms they would describe the events. Several models presented themselves. On the one hand, there was the interpretation as prodigies, along the lines of the widespread pamphlet literature that had emerged especially in Germany in the years after 1531. As explained above, the core of such pamphlets generally consisted of some barebone news report, originally probably a letter by some eyewitness or person with knowledge of the event. When printed, such reports were typically complemented by a religious interpretation. Sometimes this commentary was even longer than the actual report. Such pamphlets aimed not just to inform, but also to educate the wider public. By standards of their time, they seem to have been mass-products. Of course, their production normally had a commercial dimension: some book-seller felt they would sell. But they would never have become popular without the active support of the authorities. The whole genre was definitely promoted by the Lutheran clergy. In most cases, it must have been clergymen who not just approved the publication, but acted as authors or editors as well.[1]

On the other hand, authors could follow a philosophical model. There was still a strong humanist tradition that rejected any prodigious explanation and saw the discussion of earthquakes mostly as a learned exercise. This sceptical attitude to prodigies initially was not limited to academics and learned humanists. In Italy at least, those views were widely divulged in vernacular treatises. Dal Prete has recently drawn attention to the large number of vernacular meteorologies in Italy, which for a long time mostly ignored theological demands. He especially drew attention to the age of the earth: these authors do not follow biblical chronology, but philosophical arguments which lead to the earth being much older, if not eternal.[2] Topics like earthquakes were likewise discussed based on secular, philosophical arguments. An example is an overview of meteorological theory in the vernacular by Vitale Zuccoli, published in 1590. Zuccoli was a theologian as well as an accomplished poet and man of letters. His book has an explicit philosophical approach. The title announces that it explains the "marvellous things that are generated in the air, and some wondrous properties of sources, rivers, and the sea, according to the doctrine of Aristotle". Earthquakes are discussed according to the scholastic pattern. There is only a cursory reference to the role of God. On the question why plague seems to avoid places that have been hit by an earthquake (definitely not a standard question in earthquake treatises), the answer is: "We could attribute this to the power and judgement of God, to whom it pleases to castigate some mortals in one way, some in another. But speaking according to natural reason,...."[3]

However, with the onmarch of the Counter-Reformation in Italy, the religious interpretation would increasingly impose itself. From the second half of the sixteenth century onwards, the interpretations of theologians and Church leaders would come to set the agenda for histories and other non-religious works aimed at a general public. Vernacular authors became much more cautious. In the seventeenth century, the Bible, as interpreted by orthodox theologians, became guiding in their works.[4] This unavoidably would have its effect on the interpretation of earthquakes as well.

In the first example discussed below, the earthquake of Ferrara in 1570, this development is still underway. Most reactions are by humanistically inspired authors. Their writings have the form of dialogues and they mostly have a philosophical outlook. Still, religious elements are already sneaking in. In the reactions to earthquakes in northern and central Europe a few years later, on the other hand, the religious interpretation is completely dominant. These reactions are mostly written by Protestant authors, many of them clergy, and often take the form of a sermon.

The earthquake of Ferrara (1570)

One of the most spectacular earthquakes of the century was the earthquake swarm that hit Ferrara, at that time a sovereign Italian duchy under the house of Este, in 1570. It took several years before the quakes finally subsided, leaving the population under considerable stress. The ducal family temporarily left their palace and camped in the gardens. As common under the circumstances, religious ceremonies were organized. The Duke and his family set the example with a procession in their gardens. Carnaval in 1571 and 1572 was cancelled.[5]

The news was divulged widely, in letters and pamphlets. It is reported in the collection of Wick in Zürich. A pamphlet printed at Augsburg, the translation of an Italian letter, gives a largely factual description, with a few pious remarks.[6] A French pamphlet on the other hand emphasizes the hand of God: "There cannot be any subterfuge here, one should not adduce any natural reasons: the matter is so evident that a blind man can see it." People who seek such subterfuges are called "atheist". This earthquake is different from quakes in the past; as it damaged the dikes, it persecuted the victims even with inundations.[7]

In Italy and Ferara itself, several authors put their thoughts to paper. The Duke's secretary and court historian Giovan Battista Pigna wrote a series of sonnets on the occasion.[8] A number of vernacular treatises were published. Many more probably remained in manuscript. These treatises for the most part continue the earlier humanist discourse and are written in the form of dialogues. One such dialogue was written by Lucio Maggio and published in Bologna in 1571.[9] A French translation by Nicolas de Livre, lord of Humerolles appeared in 1575 and was reprinted in 1579. Jean Bodin contributed a poem to de Livre's book.[10]

Maggio relates how he met a friend and some other travellers amidst the ruins of Ferrara. Of course, they experience an earthquake. Together, they continue to Bologna. In the barge which carries them, one person, to pass the time, gives an explanation of the causes of earthquakes, mainly following Aristotelian philosophy. Another member of the company objects by claiming that it is rather the fire (from burning exhalations) that causes earthquakes, in the way of a siege mine. The first speaker answers that the cause is really the exhalation. Sulphurous material is converted into an exhalation by the power of fire. If this matter finally catches fire, that is *per accidens*. Other standard characteristics of earthquakes are discussed on succeeding days. Here again, first the Aristotelian theories are explained, after which other people ask questions and give comments.

The dialogue focusses almost exclusively on natural explanations. As far as religious elements are concerned, the first speaker ridicules the idea that mock suns are an evil omen and announce the death of a prince.

> As if there exists anything on earth, so excellent as it may be, the perishing of which alone would be of such consequence for the whole universe, that it would force the celestial bodies to show a particular alteration. For, as Aristotle says, the thought that because of such a tiny mutation, the heavens should be moved from their state, is far too vain.[11]

On the other hand, when it comes to the remedies against earthquakes, it is concluded that the best remedy is to have recourse to God, "who although he operates by means of secondary causes, is still the first cause, and impresses greater power into the effects than the secondary causes".[12] The author exhorts to prayers, fasts, piety, and almsgiving. He even includes some of the old miracle stories.

Another dialogue was written by Annibale Romei. Although apparently written on the occasion of the earthquake of 1570, it was published only in 1587, at Ferrara. The characters are fictitious. On the first day, Erofilo acts as teacher, Cintio as listener. This day is mostly devoted to the causes of earthquakes and meteorological phenomena in general. On the second day, they are joined by Teodoro. They then discuss the specific causes of earthquakes and some other meteorological phenomena. Here again, Aristotle's theory is abandoned for a theory of exhalations caused by subterranean fire. Other standard questions are discussed as well, without always following Aristotle. So, when it comes to the classification of earthquakes, Romei lists nine kinds.[13]

As in the medieval tradition, Erofilo is critical of portents. He almost literally repeats what Maggio wrote: Comets may have harmful effects in a natural way, and likewise earthquakes because of the excessive heat that the earth receives;

> but it is an expression of vanity to say that they are a sign of the death of a great prince, because that would require that a tiny part of the lower world would move the higher world to create marvellous effects.

Images in the sky should not be deemed miraculous, and they too depend on natural causes. The religious elements in the work are restricted to the introduction and the conclusion. At the very beginning of the first dialogue, Erofilo explains that nothing happens in the world without God's will. However, God chooses not to rule directly. He rules the heavens by means of angelic intelligences, whereas in the subterranean world, "he has given the sceptre in Nature's hand, who, aware of the Divine secrets, always aims at the good of the universe. So, earthquakes depend on nature with the concourse of the divine will".[14]

This is still well in line with the general naturalism of the work. It is only at the very end of the dialogue that some allowance is made for the supernatural. Erofilo admits that even though God has given everything its natural cause, he can make things work against their nature "to punish the sinners".[15] Still, most harm that follows earthquakes is simply natural. Finally, it is asked whether an earthquake is possible that strikes the whole Earth? Naturally speaking, this is impossible. Therefore, the fact that during Christ's death on the cross the whole earth trembled, must have been a miracle.

Still another dialogue on the earthquake was published by Jacomo Antonio Buoni, a physician from Ferrara. His work was published in 1571 at Modena. The dialogue is held on four successive days. The first three days, there are three interlocutors, all taken from real life. Apart from Buoni himself, there are Benedetto Manzuolo, a philosopher, and Alessandro Sardo, a historian. Interestingly, this very Alessandro Sardo also wrote a treatise (not a dialogue) on the earthquake himself, that he dedicated to Buoni. The work was included in a volume published in 1586 of Sardo's treatises.[16] Before looking at Buoni's dialogue, let us have a look at Sardo's treatise.

In his introduction, Sardo explained that the earthquake he experienced at Ferrara led him to an investigation of the cause, kinds, effects, signs, etc. of earthquakes. The work is indeed largely a rehearsal of these standard points, whereby Sardo refers to the various answers that have been given by other authors. Not always does he draw a clear conclusion, but as far as causes are concerned, he favours the Aristotelian explanation. In some cases, as on the duration of earthquakes, his experience leads him to divert from the received opinion.

He does discuss the final cause of earthquakes. Again, he rehearses various opinions: according to Milich, it is the signification of some future event, according to Nipho, it is the general good. In the end, Sardo does not doubt that earthquakes are divine instruments. However, he prefers to give natural explanations. As to the ancient notion that earthquakes announce war, plague, and famine, he is critical. First, he refers to the scholastic view that these effects are caused naturally; but then, he states that in many cases, these effects are not following at all. He closes with some pious remarks, emphasizing God's goodness rather than his anger: "It is certain that the divine goodness castigates, and not extirpates his children who beseach him."[17] He ends with the prayer that allegedly stopped the earthquake in Constantinople in 446, the Trisagion.

In Buoni's dialogue, these physical questions are complemented with a religious part. On the first three days, Buoni, Sardo, and Manzuolo discuss various philosophical aspects of earthquakes, they pass in review past earthquakes, etc. The third day ends with the question whether earthquakes have a prodigious nature. Buoni apparently felt this question beyond the competence of secular learning, so on the fourth day, the interlocutors are joined by a theologian, Agostino Righini. The discussion then turns towards the question whether earthquakes have natural or supernatural causes. Righini defends a Thomistic position. The earthquakes have natural causes, but nature is still subject to God, who uses it either to sustain or to punish us. A long list of passages from the Bible and other authoritative writings must demonstrate that earthquakes are actually God's instruments.[18]

At first sight, these dialogues appear almost completely secular. Although the authors do include some exhortations to piety or even pay lip-service to the doctrine of the Church that earthquakes can be supernatural warnings, they rather emphasize that most earthquakes are natural occurrences that have no special meaning. All authors are critical of the prodigious nature of earthquakes or other unusual events. If God is involved, it is as a first cause, not by working against nature. Earlier historians indeed focussed on the secular character and try to find an explanation for it. One author has explained it as resulting from political motives. At the time, the Duke of Ferrara was at odds with the Holy See and would therefore promote interpretations which denied clerical influence.[19]

Looking closer, however, the secular character needs to be qualified. As we have seen, downplaying theology appears a common tradition among sixteenth-century Italian authors, including in such staunchly Catholic centres as Naples. Compared to earlier texts, like those by Beroaldo, Porzio, or Toledo, the dialogues from 1570 strike rather for their inclusion of pious remarks. Even if one feels that these are mere lip-service or are included to keep the clergy happy, it is still relevant that at this point in time, authors felt the need to do so. The controversy between the Holy See and the Duke of Ferrara was no doubt real, but it was the former who was on the offensive, trying to impose the new Counter-Reformation view that earthquakes, instead of being mere natural occurrences, were sent by God for specific purposes.

Apart from these dialogues, a number of other works was published. The historian Pirro Ligorio wrote a treatise wherein he questioned natural explanations and stated that earthquakes should first of all be explained theologically.[20] The philosopher Gregorio Zuccoli used the opportunity to attack Aristotle's philosophy.[21] August Galesio, professor of philosophy at Bologna, wrote a philosophical treatise wherein he mostly reviewed the opinions of earlier authors.[22] There was also a work written in Hebrew by a Jewish rabbi from Ferrara, Azariah de' Rossi. Rossi knew and used Buoni's dialogue. He gives a detailed account of the earthquake, focussing on the Jewish community of the city. Throughout the narrative, he uses biblical language and traditional metaphors, but he often reinterprets the texts in the light of his own experiences, diverting from traditional Jewish exegetes.[23]

With the onmarch of the Counter-Reformation, naturalistic views had to make place for more religiously oriented interpretations. Ecclesiastical authors would take over and transform or replace this earlier discourse by an argument that claimed earthquakes were divine warnings rather than fortuitous events. Let us conclude with a work of a somewhat different character, written by a clergyman. In 1597, Simone Maiole, Bishop of Volturara and Montecorvino, published a book titled *Dies caniculares*. It was a popular work that was reprinted as late as 1654. The book is in Latin, but it is not a scholarly work. It is a work of instruction, wherein Maiole as a cleric explained how to think about nature. It consists of a number of dialogues on nature between a theologian, a knight, and a philosopher, thereby following a common model for books addressed to educated laymen at the time. A large array of topics are passed in review – humans, animals, mountains, plants, fire, stars, works of art, and so on. There is a certain predilection for the spectacular and the prodigious. Meteors get a lot of attention, especially unusual ones and meteors that should be regarded as miracles. The system of Copernicus is also referred to (and refuted).

There is no real debate among the interlocutors. They agree upon the fundamental points and just put forward different aspects. In the chapter on earthquakes, the theologian remarks that nothing is heard about earthquakes before the foundation of Rome, implying that earlier, they did not occur. According to him, God makes his wrath manifest by an earthquake only when the freedom of the Church is violated by the kings; that is, when kings attempt to usurp the domain of the sacred. He gives a long list of examples, both from ancient and from more recent history. There follow other examples of prodigious earthquakes, as the one that destroyed Atlantis or the one at the Crucifixion. Then, there are stories of big earthquakes that left only churches or houses of bishops standing undamaged.[24]

The philosopher admits that these things are truly miraculous. He adds some examples of earthquakes that are natural but still wonders. The knight points out that Rome has been largely free of earthquakes since the birth of Christ, and regards this as a consequence of the promise that "the gates of hell shall not prevail against it" (Matthew 16:18). Not just the city, the Pope too enjoys divine protection, as has been seen in an earthquake under Boniface VIII. The theologian does not object, although he does point out that Rome still has been subject to quakes provoked by human misbehaviour. In a later addition to the work (not by Maiole), it is pointed out that Rome is not immune from earthquakes, but the firmness of the true faith makes that even if it shakes, the city cannot be damaged too much.[25]

So, this book largely ignores the Aristotelian explanations and prefers to focus on the moral and religious aspects. Still, it does pick from the many examples and arguments that had been brought together in the scholarly literature. Interestingly, Maiole's book was later used as a source by some authors of learned meteorological works.[26]

The earthquake in the Streets of Dover (1580)

The first earthquake in northern and central Europe which gave rise to a sizeable number of printed reactions is the earthquake in the Streets of Dover on 6 June 1580. This quake was felt in a wide region in France, England, and the Low Countries. Most of the literature on the earthquake comes from England. A few pamphlets are known from France. In the Netherlands, there were no printed reactions, although at least one reference is known from archival sources.[27] The discrepancy may partly be due to the fact that in England, the earthquake was felt in London, with its many writers and printing houses, whereas in France it was only felt in more peripheral regions. Also, the Church of England was by now in a phase of consolidation, so that instructing the believers was a priority, whereas France and the Low Countries were still very much in the chaos of civil war.[28]

The French reactions were initially all published at Paris (where the earthquake had been hardly felt or not at all), with the permission of the (Catholic) religious authorities. The pamphlets tell us that immediately after the quake, processions were held in the afflicted cities. In Rouen, these processions are said to have been held from the beginning of the quake until midnight. The authorities demanded that shops would close and that everybody would join the prayers.[29]

In general, the French pamphlets offer only little information about the actual events and are much more concerned with the religious interpretation. One short pamphlet, reprinted at three other places, consists of a preface to the Christian reader, a short report of the quake, and an "exhortation to the Christian people to lead a good life". The report emphasizes that the earthquake comes from God, actually reusing a whole passage from the earlier pamphlet on the earthquake of Ferrara:

> There is no matter here of looking for some subterfuge, one should not adduce natural reasons: the case is so evident, that even the blind can see it. I know that the physicists will oppose what I say, as similar things have happened in the past (...) because they have read certain reasons of such events. Still, I feel sure that in the end they will be forced to grant me that we have to look differently here, and that natural reason is all in disorder.[30]

A second pamphlet, "Discours of a wonderful and true copy of the grand Deluge", consists for one half of an edifying introduction, wherein the author emphasizes the fact that God acts against the order of nature; the other half is mostly based on hearsay. Besides, the pamphlet contains a few pages of pious verse.[31]

The event gave also occasion to a more substantial work on earthquakes in general. The anonymous author, clearly a learned person, refers to numerous sources, both philosophers and historians. He rehearses the standard scholastic points, but also includes a few less common elements, such as a peroration that cities are not immortal. He lists the various philosophical opinions on earthquakes, but emphasizes that the true cause is uncertain and maybe beyond human reach. That does not impede him to come up with a theory of his own: referring to the experiments of Hero of Alexandria, he feels that earthquakes may be caused by the compression and expansion of air within the earth. In the end, earthquakes should be regarded as signs outside the ordinary course of nature, sent by God. This however does not imply that one cannot assign any natural cause to them.[32]

A short description of the 1580 quake is included. The author feels that, although the damage is much less than in the great quakes known from history, the earthquake still bears comparison to these: it happened in regions where such things had never happened before, and in marshlands, which according to the ancients should be free from such disasters. If it nevertheless appears small, "therefore we have to thank God, because he has castigated us more leniently than many other provinces, even though these may have offended Him less than we have done".[33]

In England as well, (Protestant) Church authorities did promote a religious interpretation. The archbishop of Canterbury issued a special order of prayer, to be used in all parish Churches and households, to avert and turn away God's wrath. There was also issued an admonition in the form of a homily. This did not remain a purely ecclesiastical affair, since the Privy Council,

> Considering the State of this time, wherein it hath pleased the most highest for the Amendment of all sorts of People, to visit the most Parts of this Realm with the late terrible *Earthquake*, as an extraordinary Token of his Wrath against them (...).

expressly condoned the archbishop's actions and required that he extended his order over the whole Kingdom.[34] Consequently, by order of the Privy Council an "order of prayer" was printed which included the prayers, Psalms and biblical verses to be used in Church services. It also contained a prayer "to be vsed of all housholders, with their whole familie, euery Euening before they go to bed". It asserted that the earthquake was a sure sign that the Day of Judgement was not far off.[35]

A number of writings were published in England on the quake: pamphlets, treatises, ballads. Not all of these have been preserved. Most of them have a clear religious message and interpret the earthquake as a prodigy. The minister Abraham Fleming, who had earlier written a book on the comet of 1577, now published a work on earthquakes. It was based on Nausea's 1531 treatise, though much expanded by Fleming's own comments. He also added seven extra chapters, including a warning of the Last Judgement, a list of earlier earthquakes and prodigies that had happened in England, and a prayer for repentance. Fleming saw earthquakes exclusively in religious terms.[36]

A book published in 1580 that according to the title contained "many wonderful examples of God's indignation", gave a report of what happened in England during the earthquake. The author argued that the signs that (according to Aristotelian theory) used to accompany earthquakes proceeding from natural causes were not present in the 1580 earthquake. Consequently, "this Earthquake was not natural, but of God's own determinate Purpose (…) to shew the Greatness of his glorious Power, in uttering his heavy Displeasure against Sinners".[37]

We find a similar line of thought in a treatise by the physician Thomas Twynne, who had earlier written a treatise on the comet of 1577. He admitted that he himself had not noticed anything during the earthquake and gave a description "as I haue been enformed by persons of credite".[38] He first of all gave a long explanation of Aristotelian theory. Comparing this with the things that had been observed during the earthquake, he felt, would enable one better to discern "whether it be meere naturall, or no".[39] In his conclusion, he remained rather uncommitted as to this question, maybe because the 1580 earthquake in the end was not that spectacular. This did not deter him from emphasizing that we should attribute it to God's providence and repent our sins. On the other hand, he offers a rare example of someone who sees earthquakes not just as purely negative, but feels they may have a positive meaning as well. He suggested (a "coniecture", he called it himself) that, after so many years that under Queen Elizabeth the true faith had been propagated, the Lord, as a token of his consent, had given a nod with his head, whereas, as the Bible says, "all the earth doth shake".[40]

These works very much have an edifying purpose, but even among the learned, the interpretation of earthquakes as portents made headway. The scholar Gabriel Harvey wrote an account of the earthquake in the form of a letter to his friend Edmund Spenser, a kind of dialogue again, allegedly reporting a conversation he had had when the earthquake struck. Challenged by his companions to give his opinion, Harvey gave a long explanation of the nature of earthquakes. It contains many familiar elements. As to the question of natural versus supernatural causes, Harvey cautiously balanced between the two sides of the argument. The material and formal causes are natural, but the final and efficient causes should be deemed supernatural. The final cause may be simply the wind to recover its natural place, "but sometime also, I graunt, to testifie and denounce the secrete wrathe, and indignation of God, or his sensible punishment vppon notorious malefactors…". Still, Harvey emphasized that even such moral ends were "commonly performed, by the qualifying, and conforming of nature, and natural things, to the accomplishment of his [=God's] Diuine and incomprehensible determination". In other words, he fell back to the well-established distinction between first and secondary causes.[41]

Harvey went on to talk about earthquakes as signs announcing evil, but then immediately pulled back again and stated that he dared not be so bold as to claim that he could decide that the earthquake at hand, or any earthquake, was a supernatural and immediate action of God for some specific reason, "when as I am sure, there may be a sufficient Naturall, eyther necessarie or contingent Cause in the very Earth it selfe". Without denying that God was behind the event, he stated that it would be utterly presumptuous for any man to claim that he knew God's hidden intentions, as "if they had a key for all the Lockes of Heaven".[42]

In conclusion, the English Church and government were much given to bring the people to piety and propagated the interpretation of the earthquake as a divine warning. In France as well, this interpretation dominated. Scholars might still come up with their own views, but even if they were critical of the confessional interpretation, they discussed it at length.

The earthquake of Vienna (1590)

An event that elicited a large body of interpretative literature in the German lands, comparable in size to the Italian literature of the 1570 earthquake of Ferrara, was the earthquake in Lower Austria in the night of 15–16 September 1590. Several aftershocks followed in the next days. The quakes caused damage over a wide area and were felt as far as Prague. Vienna, the Austrian capital, was particularly hard hit. There was serious damage in the city and several casualties. Fear gripped the population. Many people fled the city or stayed in the open. The Archduke and his family took recourse in a lightly built construction that was hastily built in the garden.

Here too, religious meetings and ceremonies were organized. There was a prayer-meeting on 21 September, the feast of St. Matthew, in the church of the Jesuits. The bishop of Vienna, Johann Caspar, held the sermon. Two days later, major processions and religious ceremonies were held in the city. About a month later, on 19 October, the bishop held another sermon in the church of the Augustinians, again at the occasion of a procession. The two sermons were printed together in 1591 as two "Catholic sermons, held in public assemblies for common prayer against the terrible earthquake".

In his sermons, Caspar discerned three kinds of earthquakes. Two of these were only metaphorical, to wit, earthquakes in human life, and spiritual earthquakes, or the motion of our heart. As for quakes in the literal sense, to which he allotted most space, Caspar did not deny that these had natural causes, but emphasized that God was the first cause and could use them for his ends. God's main aim with earthquakes appeared to have been to punish humans. From biblical and profane history, Caspar gave a long list of examples of earthquakes sent for this purpose. So, the earthquake at the crucifixion demonstrated the great sin of the Jews in crucifying Christ. Humans were especially punished because of their refusal to do penance for their sins. Interestingly, Caspar referred to the Stoic idea that one should vanquish the fear of earthquakes by despising death. He admitted that the idea was good in itself, but needed a supplement: death was harmless only as long as one did not die in state of sin. One should not fear natural death, but the death of the soul was the supreme danger.[43]

This earthquake drew wide attention. News sheets and pamphlets were published in various cities in Germany. One such pamphlet was printed in Cologne, a Catholic (but rather liberal) city on the Lower Rhine. Another one was printed at Frankfurt on the Oder, in far-away (and Protestant) Brandenburg. Both appear to go back to eyewitness reports. The Cologne pamphlet gives a factual chronicle of the quakes and describes the damage done in Vienna, especially to churches, as well as the effects on the population. The main part of the Frankfurt pamphlet appears to have been written by someone in Prague. The author reports on how the earthquake had been felt in Bohemia. This text, of about five pages, is then followed by a much shorter (two pages) report of what had happened in Vienna, clearly written by someone else. The publisher simply put two reports together to come up with a publishable work.[44]

In both pamphlets, the factual narrative is put in a religious framework. The Cologne pamphlet has the report preceded by a long preface, explaining that God sends signs to warn us to mend our evil ways. The pamphlet concludes with a prayer. The Frankfurt pamphlet, on the other hand, does not need such an introduction, as the report itself has a clear religious tenor. The factual information contained in the report from Prague is rather short and most of the author's attention goes to the meaning of the earthquake as a divine warning, referring to biblical and historical precedents. (The short report from Vienna which has been added, on the other hand, is factual again.) In both cases, the pious message is emphasized by the paratext. Both titles explicitly refer to God's wrath. The illustration on the title page in the one case shows the Last Judgement with the dead rising from their tombs, in the other the hand of God with a rod and a burning church in the background.

It is striking that, contrary to the earlier reactions in Italy and even to a certain extent in England, the humanist, more secular voice is completely absent. Clearly, the intellectual and religious climate did not encourage to bring such views forward, but of course it is unlikely that such people were not around at all. The papal nuncio in Prague reported that people saw the cause of the event in the heavy draught of that year, although he added that the wiser people attributed it to the sole will of God.[45] In a later work, the physician Hippolytus Guarinoni tells of an oral dispute on the significance of earthquakes that his father, then an imperial physician at the court of Vienna, around 1593 had with another imperial physician. The other, according to Guarinoni, laughed at the notion that earthquakes had a religious significance and maintained that they were caused by mere nature. Guarinoni's father emphasized that they were divine instruments. To Guarinoni's satisfaction, his father was blessed with a long healthy life, while the other died young.[46]

The earthquake occurred at a critical moment in the relations between Catholics and Protestants in Austria, with the Protestants put ever more on the defensive. This may explain why, although Vienna was Catholic territory, most of the reactions that have come down to us are by Protestants. The main Catholic reaction, apart from Caspar's sermons, is an earthquake chronicle by the mathematician Johann Rasch, who had earlier published a work on earthquakes in 1582.[47] Most Protestant reactions are sermons; it is with the earthquake of Vienna that the big stream of printed earthquake sermons starts. Besides, there is a *Newe Zeitung* on the earthquake, published by the minister Marcus Volmarius in 1591. "Newe Zeitung", literally something like "new report", is normally the title of a news sheet, but in this case, it is not a factual report. It is rather a theological exegesis of the *Zeitung* that had been published earlier, wherein Volmarius explained the significance of the various events that had been reported. They all address certain specific sins and sinners.[48]

Most prominent among the Protestant authors is Konrad Kircher. He had earlier been a Lutheran minister in the imperial city of Augsburg, but Lutheranism had been suppressed there. At the time of the earthquake, he was a minister at Sonnberg, a village some twenty kilometres north of Linz. After the quake, he held a series of no less than ten sermons on this earthquake. They were published only in 1594, after he had moved to Donauwörth in Germany. Printing was encouraged and probably supported by a local noblewoman, who had read a copy of them and to whom the sermons were dedicated.[49] The evangelical minister David Schweitzer held a *Christlicher Busspredigt* [Christian sermon of penitence] at Schöngraben, some 50 kilometers north of Vienna, the 14th Sunday after Trinitatis, 1590. There is a dedication by Schweizer to some Austrian notables, dated 9 September 1591, in which he said that he had agreed to print it, but the work was printed at Frankfurt only in 1593 by some other nobleman who had obtained a copy from the author. This nobleman's preface is dated August 4, 1592. Apparently, sermons like these circulated in manuscript; probably, only a tiny portion of them made it into print.[50] Still another "Christlicher Busspredigt", held the second advent Sunday 1590 by the minister Johannes Phyldius at Oberursel (near Frankfurt), was equally printed in 1593. As Phyldius explained in the dedication, many had read and copied his original sermon and to this day, he was asked to have it printed.[51]

These sermons regard the earthquakes strictly from a biblical point of view, although in some cases the problems of the day do transpire as well. Some prefaces explicitly refer to the threat of a new war with the Turks. Kircher in his preface pointed out that in the past, earthquakes had often announced a Turkish invasion, and he suggested that the recent earthquakes might announce a new war. Kircher does not go so far that he rejects all natural explanations of earthquakes, but he feels that these are in the end insufficient. In the end, earthquakes are a divine miracle.

> Now, because it is a divine miracle, reason cannot comprehend, still less describe it. If reason could comprehend and describe it, either man were God (…), or God were a man, who could do nothing higher in nature than what we humans can understand and comprehend by our reason.

So, to know what is an earthquake, one should have recourse to Scripture. An earthquake "is a step of the almighty God which He makes on earth in His anger, when he comes down. Thereof mountains and valleys tremble and quake".[52] Kircher referred to a number of biblical sentences, although he admitted that there were exegetical subtleties. In the case of incomprehensible miracles, the Holy Ghost spoke metaphorically, so probably one should not take the definition too literally. But:

> If a father in dismay and anger walks around in the room and stamps so forcefully on the floor, that the windows tremble, the child shall not inquire why it is that the windows tremble that way. He will easily see that the father does it with his angry steps.[53]

Earthquakes were worked by God because of His anger. They announced impending great things, in particular, the Day of Judgement. Therefore, for true believers, they were a source of solace as well. Kircher spoke at length about how we as Christians should prepare for the impending end.

The sermon by Schweitzer comes in three parts. In the first part, he discussed the cause of earthquakes. There are various opinions, some of which should be abhorred by all true Christians.

> For some inconsiderate people, as the godless among the common riffraff, are of the opinion that earthquakes proceed only from the ordinary course of nature. They say: it has from the origin of the world generally been the case, that at times the earth has trembled and moved. Also, they have heard from their ancestors that similar earthquakes have occurred years ago, and that these were followed by good years, fertile in crops and cattle. Consequently, that one hears of them again would be nothing new.[54]

These people commit a sin and deny divine providence, for the Bible teaches that nothing in the world happens without a purpose.

Schweizer discussed the opinions of the astrologers and of Aristotle. As for Aristotle, he admitted that his views were not to be rejected and that earthquakes did have natural adjunct causes (*Neben-Ursachen*), but this was not the main thing. In themselves, such causes are powerless. Only when God gives His assent and shows his omnipotence, the work gets done. God himself is the first and true efficient cause. Yes, there are malign planets and putrid winds and vapours that cause earthquakes: these are the world's sins, wickedness, and vice. Although Schweitzer basically rejected learned philosophical opinions, he still treated them with respect. However, he had no patience at all with opinions that allegedly existed among the common people, that saw earthquakes merely as natural events and as harbingers of fertile years.

The second part of the sermon explained, with many examples, that earthquakes announce impending evil, by which God wants to punish our sins. Earthquakes could mean a change in religion. They announce the Last Judgement. God made his entry as an earthly prince, preceded by heralds (signs and wonders) and gunfire salutes (earthquakes). The third part spoke on the necessity of doing penance.

Phyldius ordered his sermon according to the topics: what are earthquakes, how do they originate, what is their meaning, how can they serve us. He first gave a chronological list and a description of earthquakes. As for the cause, he emphasized that nothing in the world happened by chance and that earthquakes were ordained by God, who caused them both with and without natural causes. The causes given by the philosophers might be true, but these do not take away from the government and omnipotence of God, who had ordered the world this way. As for meaning, Phyldius gave a long list, corroborated by historical examples, of the various sins which might occasion earthquakes: idolatry, blasphemy, persecution of ministers, rebelliousness, and disobedience. He gave a list of historical examples of earthquakes occasioned by them. Especially persecutions get a lot of attention. This may reflect the precarious situation of Protestants in many places at the time. Phyldius closed by emphasizing the necessity of doing penance and by explaining that God's mercy can save people even from the most pressing danger.

There were other reactions as well. At the Protestant academy of Helmstedt, the earthquake was the subject of an academic oration by the professor Johannes Heidenreich, printed in 1591. He claimed that the earthquake could not sufficiently be explained from physical causes and therefore had to be supernatural. It was a clear divine warning. Heidenreich also wrote some Latin poetry on the event, as did other professors.[55] On a more popular level, the earthquake was referred to in *Meistergesang*. *Meistersinger* were craft-like organizations in the German lands for the practice of vernacular poetry. Most members originated from the middle classes. Their poetry followed certain set rules. Most of it remained unprinted, but many manuscripts have been preserved. In the sixteenth century, most of the *Meistersinger* became part of the Reformation movement and put their poetry in the service of the Protestant cause. Many of the poems, or rather songs, had a didactic and moral tenor. Some of them refer to portents, including earthquakes that punish sinners.[56]

From the *Singschule* (Singer's guild) of Iglau (present day Jihlava in the Czech Republic, halfway between Vienna and Prague), there is preserved a manuscript with eight poems, all titled "on the earthquake" (*Von dem Erdbeben*). The manuscript and two of the songs are incomplete, so there may have been more. The songs are anonymous. One is dated 11 November 1590, four others on the first four days of February 1591, the remainder are undated. It is clear that they have been composed under the impression of the Vienna earthquake (which was felt in Iglau as well, though less severely), but in only two of the poems there is explicit reference to it. The others rather list earthquakes that happened in the more distant past, back to antiquity, with the dire consequences that followed them. Some explicitly put this in an eschatological context and explain that God sends warning so that people may repent in time. Some leave this message implicit. In all cases, it is the portentous meaning of the earthquakes that counts.[57]

Comparing the reactions to the earthquake of Vienna or the Streets of Dover to the one in Ferrara twenty years earlier, we obviously have to take into account the difference in genres. A sermon is a different genre with a different message than a dialogue or a work of history. However, the choice of genre is in itself significant. In Germany, most of the public reactions on the earthquake are by Lutheran ministers, whereas in Italy, they were mostly written by secular intellectuals trained in the humanities. The Italian works above all wanted to instruct in an entertaining way, whereas the German reactions were part of an active campaign to propagate piety and turn people into obedient Church members. This attests to the emergence of a new intellectual climate, the result of a powerful social and political change that would soon engulf the Catholic world as well.

Notes

1 Leppin (1999) 51–52 found that two-thirds of the authors of apocalyptic pamphlets were Lutheran clergy, many of them high-ranking.
2 Dal Prete (2014).
3 Zuccolo (1590) 95.
4 Dal Prete (2014), *passim*.
5 Solerti (1891) xci–c reproduces letters by the Florentine ambassador and others on the events.
6 *Ausszug* (1570). Wick (1975) 184–185.
7 *Discours* (1570) 4, 12.
8 Solerti (1891) xcix.
9 Maggio (1571). For a discussion, see Martin (2011) 70–71.
10 Maggio (1575). Bodin's poem is reproduced and analyzed by Blair (1992).
11 Maggio (1571) fol. 13. The French translation (Maggio (1575) 17) is even more explicit: "For it is a great stupidity and worthy of ridicule (says Aristotle) to think that because of such a small change as the death of a prince, the heavens should receive some impediment and alteration in their ordinary course."
12 Maggio (1571) 49v.
13 Romei (1587) 82. On this dialogue see Martin (2011) 73.
14 Romei (1587) 35, 2 (quotes), 46 (images in the sky).
15 Romei (1587) 103.
16 Sardo (1586) 169–207.
17 Sardo (1586) 207; see also 175–176 (various opinions), 205–207 (effects).
18 Buoni (1571). For a discussion, see Martin (2011) 66–70.

19 Martin (2011) 61–62.
20 Martin (2011) 74–76.
21 Martin (2011) 74.
22 Galesio (1571).
23 De Rossi's treatise is discussed by Weinberg (1991).
24 Maiole (1654) 22–25.
25 Maiole (1654) 306.
26 Resta (1643) 494, 504.
27 It was referred to in a conversation among neighbours in the city of Woerden. See Frijhoff (1995) 383.
28 On the earthquake, Neilson et al. (1984); Haynes (1979); Campbell (1940–1941); a short discussion of some reactions in Kocher (1953) 111–114. See also Ockenden's preface in Twynne (1936).
29 *Discours merveilleux* (1580) Aiij r+v.
30 *Discours merveilleux* (1580) Aiij. Cf. *Discours* (1570) 4–5.
31 *Discours d'une merveilleuse et veritable copie* (1580).
32 V.A.D.L.C., *Des causes et effects* (1580) 27–30 (cities not immortal), 5–6v (causes are uncertain), 11v–13v (compressed air), 5+v, 30+v (religious meaning).
33 V.A.D.L.C., *Des causes et effects* (1580) 25. The description of the 1580 earthquake is on fol. 24v–25v.
34 Strype (1725) II, 668–669; Strype (1710) 247–249, quote on p. 248.
35 *Order of prayer* (1580), C4v.
36 Fleming (1580).
37 Quoted from Strype (1725) II, 72, who gives a summary of the book. No copy seems to have been preserved.
38 Twynne (1936) 24. The treatise was published in 1580 under the title: *A shorte and pithie discourse, concerning the engendring, tokens, and effects of all earthquakes....*
39 Twynne (1936) 24.
40 Twynne (1936) 36–37.
41 Harvey (1966) 454. Cf. Haynes (1979) 543–544. Haynes seems to accept Harvey's text at face value as a personal account.
42 Harvey (1966) 455.
43 Caspar (1591) 47v, 48v.
44 *Warhafftige und eigentliche Beschreibung* (1590); *Kurtzer bericht* (1590).
45 Text in Gutdeutsch et al. (1987) 114.
46 Guarinoni (1616) 118.
47 Günther (1890) 243–244.
48 Volmarius (1591).
49 Kircher (1594).
50 Schweitzer (1593).
51 Phyldius (1593).
52 Kircher (1594) Fij.
53 Kircher (1594) Fiij.
54 Schweitzer (1593) 33, cf. 77–78; also printed in Gutdeutsch et al. (1987) 145. For the view that earthquakes announced fertile years, see below, pp. 150–152.
55 Heidenreich (1591) both titles; Maior (1591). Heidenreich's oration has also been printed, with German translation, in Gutdeutsch et al. (1987) 123–129.
56 E.g. Brunner and Wachinger (1986–2009) XIII, 313, 454.
57 Streiz (1958) 11–14, 186–199; Brunner and Wachinger (1986–2009) I, 181 and VI, 262–265.

10

INTERPRETING EARTHQUAKES IN THE SEVENTEENTH CENTURY

Philosophers discussed earthquakes mostly on an abstract level. The explanation of actual earthquakes however was not an academic affair, but concerned everybody. The purpose was to inform and reassure the public at large. An active campaign was waged to educate the common people, both in matters of Church doctrine and in general piety. Lutheran ministers, Jesuits, and others not just discussed the issue on an academic level, but propagated their views from the pulpit and in the new print media. This required a different register. So, how were the philosophical insights put to use? We already discussed some examples from the end of the sixteenth century. In the following, I will explain the literature on earthquakes as it developed in the seventeenth century. Since it would take too long, and be tedious, to explain every author individually, I will focus on some of the main themes that they have in common. I will divide the material into two groups, Catholics and Protestants, which roughly coincide with southern and northern Europe. We are not talking about two homogeneous blocks, however. In particular, German or Swiss Catholics may share some characteristics with their Protestant neighbours rather than with their co-religionaries across the Alps.

Protestants: sermons and edifying treatises

In the seventeenth century, events in Italy and other foreign countries would remain a topic of interest in Protestant literature, but earthquakes in northern Europe itself, even minor ones, also elicited a reaction. Apart from the commentary in pamphlets, the forms that became most common in the Protestant world to comment on earthquakes were the sermon and the edifying treatise. Whereas pamphlets are news-sheets that often relate events farther away, sermons are almost always occasioned by local events. After major or minor earthquakes, like after other disasters or frightening events, ministers and pastors often used the occasion to preach upon the event. This tradition continued well into the eighteenth century. Sometimes, they sent pastoral letters. After the London earthquakes of 1750, the Bishop of London sent a letter to the people of his diocese, wherein he denounced the blasphemy and depravity of his days and called people to repentance. It was reported that ten thousand copies were sold in two days. It was also reprinted and sold in the British colonies.[1]

Most sermons, as oral productions, probably completely escape our notice, but some of them were printed. In this way, the ministers could reach a wider audience then just in church. The reason why people bought the printed version is less clear. Maybe because they needed spiritual comfort, maybe just because they wanted a souvenir from the event. Printers might be interested simply because earthquake sermons would sell better than standard sermons. Like pamphlets, sermons were public statements, and preferably held in the first days after the event, when the audience could be expected to be most susceptible to their message.

With regard to content as well, there are similarities. As we have seen, pamphlets generally had a theological interpretation. It is likely that most of the editors of the pamphlets, who put in the edifying conclusions and other passages, were ministers or theologians. However, for news sheets, the factual account was generally the heart of the matter. The edifying passages were added afterwards. For sermons, one might expect the reverse. The minister's main aim was to explain the meaning of the earthquake. Only in order to give his message more weight, especially when the sermon was prepared for printing, he might decide to include historical and philosophical details.

The distinction between printed sermons and other edifying literature is somewhat blurred. In Protestant countries, most treatises written on earthquakes are by clergy. Sometimes, local ministers felt urged to bring their admonitions on the occasion of earthquakes directly into print, without (at least, as far as we can tell) first preaching on them from the pulpit. The text by Sigismund Suevus on the occasion of an earthquake at Vienna in 1581 does not clearly identify itself as a sermon. The works by the ministers Nicolaus Höpffner (1691) and Michael Babst (1599) are called *Beschreibung* (description) and *Tractat* (treatise), respectively. Such edifying treatises are very similar in content to printed sermons and for most purposes they can be dealt with together. Treatises on earthquakes by ministers without a concrete occasion are rare.

The distribution of earthquake sermons is uneven. Apart from the occurrence of the earthquakes themselves, not all denominations appeared equally susceptible. By far the most printed sermons on earthquakes come from the Lutheran world, where theologians took great interest in prodigies and where there existed a long-standing tradition of sermons on thunderstorms (*Wetterpredigten*) and other natural events. The biblical theme of God moving or shaking the earth was frequently referred to in church carols as well.[2] Some earthquake sermons came from the Anglican world. Among the Reformed, the reactions are often more muted, although there are examples from Switzerland and Württemberg.

The message of the sermons and treatises is invariably the same: the earthquake is a sign of God's anger. God's anger is justified, because we are sinners. We have to repent, do penance, and lead a God-fearing life, lest a larger evil will strike us. This message is generally reinforced by including a list of historical examples and the familiar miracle stories. A recurrent theme is the refutation of philosophical naturalism. As we have seen, humanist scholars were critical about earthquakes as portents and forebodes of evil. Theologians on the other hand were eager to demonstrate that many phenomena were worked directly by God and that nature could only be understood within the context of the divine plan. Earthquakes were portents, not because of some natural effect or some secret interconnectedness of things, but as direct manifestations of God. Since the Reformation, the theologians had tried to impose this view on academic teaching and on society as a whole. It is not surprising, therefore, that the question whether an earthquake should be deemed natural or supernatural frequently turns up in treatises and sermons.

Authors of edifying works certainly knew about the common meteorological theory, as they raised the issue repeatedly, and took care to treat it with respect. However, in the end, most of them rejected it outright. We already saw this in the sermons after the earthquake of Vienna. Kircher, for instance, did explain the Aristotelian theory of earthquakes, making clear that he would hinder nobody who wanted to entertain himself with speculations of this kind. But he warned that such sophistries could not really explain earthquakes. Earthquakes are real divine miracles and our reason cannot describe nor comprehend them.[3]

Michael Babst, on the occasion of an earthquake at Freiberg in 1598, stated that the Aristotelian explanation is "a clever philosophical invention, and it may be allowed in case just a village, a town, or a small part of the earth trembles and is moved". However, such is not believable of large quakes like those in Ferrara or Vienna. Here we should not speak physically, but theologically. "Not to mention those miracles [*Wunderwercke*] which go completely against Physica." As examples of the latter, Babst mentions a prodigy from 753, when allegedly a prophesying mule emerged from the earth; the earthquake of Villach in 1348 when, according to reports, fifty people turned into salt pillars; and the earthquake at the death of the Saviour. Note that more recent events are regarded as on the same footing as the biblical miracle.[4]

The natural causes are also qualified in a book of historical examples, published by two Lutheran pastors in 1598 at Leipzig.

> Although the philosophers show natural causes as the origin of earthquakes, and attribute them to the stars and the effects of the planets. However, such causes are not sufficient nor perfect, for earthquakes do not happen by accident, but by God's will and decision [*Verhengniss*], and always signify some future disaster. (...) For since the earth is so heavy that it is immovable, as the 104th Psalm testifies, an earthquake must definitely signify something big and horrible.[5]

Another example is a sermon by Andreas Osiander on the occasion of the earthquake of Unterwalden in 1601. He too started by explaining the natural causes quoted by the philosophers. Such causes can be accepted as long as we realise that God is not constrained by them.

> Many earthquakes are such that one must recognise that they did not have their origins *ex dispositione naturali* (from natural causes and occasions), but proceed *immediatè* (directly) from God, as tangible *specimina* and proofs of his infinite Divine power and glory.[6]

Not all theologians are so outspoken. The reformed scholar Claude Aubéry warned that just as it is an error to feel that things happen without God's will, it is likewise an error to ignore natural causes altogether. "We should not listen to those who do not allow any mention of natural causes, and who call all those who look for reasons in natural causes, atheists." But it is telling that he felt this warning necessary.[7]

Of course, the disagreement between the sermons and the meteorological textbooks in part just reflects their various roles: edifying the community in the one case, training in philosophical argument in the other. However, the disagreement goes further than that: the ministers do expressly reject the philosophical views. Many of them will have had their training at a university. This raises the question how the actual teaching related to the content of the textbooks. The fact that the textbooks duly list the natural explanations according to the classical authorities does not necessarily imply that the authors or teachers agreed with those. In the practice of preaching anyway, the biblical example clearly outweighed philosophical understanding.

While on the one hand refuting naturalism, the ministers on the other hand were also fighting against the view that the whole world was a big *exemplum*. In medieval legends (as well as in Biblical history), acts of God generally served particular ends: to save or punish a particular person, demonstrate the saintliness of a particular saint, and the like. The view of prodigies that came to dominate the era of the Reformation was more restricted. God sent his signs to warn people, but one should not endeavour to try to read God's detailed plans from them. The events did not mean anything in themselves, they just reminded everybody that they were sinners in the eyes of God and that but for God's mercy, we are all doomed. So, earthquakes announced God's judgement, but only in a general way. Instead of giving a specific interpretation, Protestant ministers drew a more general theological, even timeless lesson from the event.

The danger that earthquakes would be seen as direct manifestations of God's displeasure with specific people or behaviours was particularly acute in cases like Yvorne or Piuro, where complete villages had been wiped out. Whereas in normal earthquakes, destruction was random and even when many people died, others survived, these disasters struck indiscriminately. Everybody in the said villages was killed in an instant, without even having had time to repent. Many people apparently wondered what these villagers had done to deserve such a fate. In the case of the disaster of Piuro, indeed, some of the reports emphasize the godless behaviour of the inhabitants before the event.

Protestant ministers would regard this as an unacceptable explanation. Portents served to call people to penitence, not to give them the feeling that they were better than others. To this, they put up the biblical example of the tower of Siloa (Luke 13:4–5): "Or those eighteen, upon whom the tower of Siloam fell, and slew them, think ye that they were sinners above all men that dwelt in Jerusalem? I tell you, Nay: but, except ye repent, you shall all likewise perish." Claude Aubéry invoked this text at the occasion of the disaster of Yvorne. He emphasized that the fact that the people of Yvorne had been punished for their sins does not allow us to condemn them.

> You ask, whether all those who were smothered and buried by the earthquake, are destined for eternal perdition? Such a temerous judgment does not befit us. For the memory of our brothers from Yvorne is also commended by piety, humanity, and hospitality. Although it is not without reason that God wanted them to be buried and smothered.[8]

Lutherans and Anglicans would refer to the episode of the tower of Siloa to the same effect.[9]

It should be admitted, however, that in practice many ministers, though acknowledging the official doctrine, followed it somewhat half-heartedly. When it concerned earthquakes far away or among a different confession, they often were less reticent. Interpretations that appeared to support their own view of the world or the position of the Church were hard to reject. Damage to churches, palaces, or the buildings of the mighty was of course a perfectly natural consequence of a heavy quake, but often seems to have been regarded as a special warning to the owners. Churches stand for more than just a building. After the earthquake in Vienna in 1590, it was the damage to the church towers that drew most attention. Some Protestant authors clearly hinted that what happened to the spires was a sign of what was going to happen to the other high positions within the empire.[10]

When sermons and treatises went into details of the earthquake under review, this concerned mostly elements that were deemed significant from a theological perspective. So, a treatise on an earthquake at Nuremberg in 1670 related that right before the shock, about two hours before daybreak, in several houses, people noticed a knocking on the doors, as if someone wanted to enter, as well as the sound of people walking in the house. In some houses, doors were opened. "Now, if God caused such a frightening sound, as if thieves at night were committing burglary, this definitely did not happen by accident, from a frightened imagination, but is rather another fatherly admonition by God."[11]

To some extent, such "omens" were probably a rhetorical ploy. Ministers in their sermons would often explain God's work by comparing it to well-known things or events. Such metaphors were not necessarily intended to be taken literally. However, metaphorical thinking was a tool to understand reality. The confessionalized disenchantment of the world did not immediately lead to a modern empirical view, but to what Bill Ashworth called an "emblematic world-view", whereby concrete things or events represented higher, general truths.[12] Moreover, the aforesaid signs were not just emphasized in sermons and printed pamphlets, but also privately written down. In 1692, a widely felt earthquake happened in north-western Europe. At Leiden, an anonymous person made a note of this event on the fly-leaves of a Bible, stating "that the bells have sounded [...] and on the anatomy room the skeletons started to move and the houses to shake".[13] The shaking of houses appears to have made less of an impression than the sounding of bells and the moving of the skeletons, which of course (though perfectly natural) could be seen as anticipating the Last Judgement. The story of the moving skeletons at Leiden travelled widely. It also turns up in a contemporaneous German chronicle, which reports that the skeletons frightened some visitors who happened to be present.[14]

In many cases, the omens were really considered as such. Ministers eagerly scanned the records and the circumstances for any indication of the supernatural character of the event. As explained before, earthquakes were often accompanied by prophecies and predictions. Educated authors were often somewhat at a loss what to think of these, but they did not reject them out of hand. Many of them were told and retold by learned authors for their moral or religious tenor. Likewise, stories of special warnings and signs were not just rumours in the streets, but also spread by ministers and priests.

The disaster of Piuro in 1618 was reportedly accompanied by many such signs. These were among others discussed by the Dutch reformed minister Willem Baudartius, author of a running chronicle of his time. The first edition covered the years 1612–1620, a second edition expanded the story to the period from 1606 to 1624. In his preface, he expressly mentioned his purpose to praise God for the protection of his Church. Among the facts that Baudartius thought worth recording, prophecies, visions, and prodigies, reported from all over Europe, figure prominently. The last volume even has a separate index for the various warnings, discussed in the book, sent by God to kingdoms and countries.

In his tenth book, Baudartius gave a detailed description of the disaster of Piuro. He returned to the event in book XI, as he now had obtained additional information from the *Messrelation*, the annual overview of past events published at the Frankfurt book fair. The new information is not about the event itself, but concerns exclusively the many signs that allegedly had accompanied the disaster. Baudartius listed them eagerly: the day before the disaster, there had been a nasty stench, nobody knew from where. An artisan warned the Chancellor Certeman and some others that they should pack their belongings and flee from the place. "But he was laughed at and ridiculed, so he left alone, saving his life." A merchant was warned by some women, but he laughed at them, and was killed as soon as he entered the town. When after the disaster the authorities wanted to dig up the place, the labourers were chased away by a horrible stench. The foreman saw a spectre that warned him to abstain from the work, as it would be in vain. An inscription in Hebrew characters was found, proclaiming the wrath of the Lord. All this is not written by a philosopher who looks down on such superstitions, but by a minister who includes them as a serious admonition to his readers.[15]

In emphasizing the miraculous nature of earthquakes, these ministers at first sight seem not far removed from radicals like Luther or Bodin. However, even if they qualify the philosophical explanations, they do treat them with respect. They do not deny that ordinarily, the world is governed by natural causes. Earthquakes and other prodigies are special events, infractions upon an established order. They are not the way that the world is normally governed, but serve to give a special message. Portents are real, but they should not be interpreted as *exempla*, relating to the affairs of this world. They are there to remind people of the larger Biblical truths. Paradoxically, it would seem that the interpretation of earthquakes in terms of divine miracles contributed to the dissolution of the medieval magical world-view.

Catholics: processions and histories

Whereas in the Lutheran world, the standard reaction to earthquakes was to keep sermons, in the Catholic world, the default reaction was to hold large public ceremonies. The most important part was generally a large procession wherein the local relics and other sacred objects were carried around. They were headed by the secular and religious authorities of the city and other prominent people, and all the social classes took part. This could be accompanied by other penitentiary measures, such as alms, lents, and the cancelling of fairs and public festivals. As noted above, large processions were held after the earthquake of 1580 in several French cities and after the earthquake of 1590 in Vienna. The tradition was however much older. It is mentioned by ancient historians and has also been recorded for many medieval earthquakes.[16]

Such processions would remain a standard reaction to earthquakes in the era of the Counter-Reformation. During the eruptions of Etna, which would happen every twenty years or so, the clergy of the nearby city of Catania would bring out the relics of the local Saint Januarius, in particular an ampule with his blood, and carry it through the city. After a severe earthquake in Málaga in Spain in 1680, the local bishop wrote a long pastoral letter wherein he pointed out that the earthquake was a result of people's sins. He called upon them to do penance and ordered processions to be held the following Sunday.[17] At Lima, processions were held not just after local earthquakes, but also after earthquakes in other cities in the viceroyalty of Peru.[18] After the earthquake of Sicily in 1693, processions and rituals were ordained by the authorities in Rome and Madrid.[19]

It goes without saying that such rituals were organized from above and hardly the spontaneous outburst of popular feeling. So it is expressly stated that after the earthquake of Palermo in 1726, the archbishop ordered these things in order to work a spiritual reformation, now that the moment appeared favourable.[20]

Especially in southern European countries, the clergy would take the lead in ostentatious ritual. The reports speak of priests girded with cords, ash on their heads, crowned with thorns, walking around in the city singing the penitential psalms.[21] Venerated relics and images would be taken out of their shrines and put on public display.

Processions in the first place were intended to fend of the Lord's anger, but they also served to bring spiritual solace and to express a sense of community. They also underlined the guiding role of the Church. Reports of earthquake often dwell at length on such processions to demonstrate the turn to piety of the population. So, Ruffo's description of the earthquake at Palermo in 1726 has a long description of the processions, listing all their participants. The German translator felt it a bit much, but because he wanted to give an accurate rendering of the original, he still included it in full.[22]

Pamphlets and news-sheets appeared in the Catholic world as well. Most of these contain a religious message. A pamphlet on an earthquake in central France devotes the first seven pages to prodigies generally, warning that they are not natural, but acts of God; the next seven pages are devoted to a description of the earthquake itself, including the processions to which it gave occasion; the last three pages are a general warning.[23] A number of popular treatises were published on the occasion of the earthquake of Malaga in 1680, most of them anonymously. All these writings emphasized that the earthquake was a punishment of God for the sins of the people. In one case, the author acknowledged that there were natural causes for earthquakes in general, but that these did not apply in this case. For the sake of piety, it was sufficient to know that they were ultimately caused by God.[24]

Although playing a less central role in the Catholic religion than in Protestantism, sermons too remained an important means of communicating with the general population. But unlike in the Protestant world, Catholic sermons were seldom printed. The two sermons, discussed above, by the local bishop, Johann Caspar, on the occasion of the earthquake of Vienna of 1590 are a rare exception.[25] A Latin oration was published by Johannes Stalenus, a priest at Rees on the Lower Rhine, on the occasion of the earthquake of 4 April 1640. It is largely devoted to the relation between natural and supernatural causes, refuting those who attribute too much importance to the former. Stalenus relies much on the Church father Saint Ephrem. In his discussion of natural causes, he is mostly concerned with the effects of earthquakes that are extraordinary – continents that become separated or founder, mountains that are displaced, etc. He does not deny that earthquakes have natural causes, but in the first place, they are the instruments of God's special providence. God uses them for the punishment of sin, in a clear show of his power, and to bring humans to humility. Moreover, they are often announcements of impending evils.[26]

However, although we do not have many Catholic sermons themselves, we do have some works that, although written in Latin, were obviously not meant for academic teaching or scholarly debate, but to provide the curators of souls with material to edify their flocks. These works too were often written by Jesuits. One author very active in writing works on general piety based on the learned discourses of the scholars was the German Jesuit, Georg Stengel. Among his many publications, in 1647 he published a book "on monsters and monstrosities that the wonderful, good and righteous God shows in administering the world". Stengel paid some attention to natural causes, but most of his discourse is on the various moral causes of monstrosities: they are signs by which God warns us; they are punishments; sometimes, they help preserve virtue and chastity; and in some cases, they are worked by the devil to terrify humans.[27]

Four years later, Stengel published four hefty volumes on "the divine sentences [*iudiciis*], that God pronounces in this world". The work was clearly intended for pastoral use. So, it contains an index where the reader can find subjects that are appropriate for preaching on a certain Sunday of the liturgical year. Three chapters, about forty pages, are devoted to earthquakes.[28] Stengel regarded earthquakes as "monsters of the earth", a point he had already made in his earlier book on monsters.[29] Although he admitted that earthquakes had natural causes, he felt that they should be regarded as monstrous because they deviated from the common disposition according to their species.

Stengel started by explaining the various ways in which God could make the earth useless for us because of our sins. In the fourth paragraph, he moved on to earthquakes. From the outset, he pointed out that he would not discuss their natural causes. The most important cause why the earth shakes is God's anger. Anger is a cause of movement, as he showed in parallel cases. The earth not only shakes, it even opens itself to devour sinners who provoke God to an extraordinary punishment. "And this is the second cause, which precedes the first mentioned. The punishment follows the guilt".[30] There followed a long list of examples, starting with the biblical Korah, Dathan, and Abiram.

Stengel continued with various other effects of earthquakes. He admitted that they might be useful as well. Their greatest use is obviously that they inspire the fear of God. Following a long Catholic tradition, he saw earthquakes happen especially in times of heresy. He referred to the natural signs of earthquakes, but then immediately went on to supernatural predictions, rehearsing some stories of earthquakes prophesied by saints who had been warned by God. Stengel included other miracle stories as well. He tells about the earthquake of Constantinople of 1509; of the earthquake at Jerusalem under Julian the Apostate (following Nikephoros' history); of the earthquake at Antioch in 528, that only ended when people wrote the words "Christus nobiscum, state" on public places; and several stories of earthquakes at a saint's martyrium. However, he also included a description of the large earthquake of Calabria of 1638, written by a fellow-Jesuit, Julius Caesar Recupito, an eyewitness to the event, that had earlier been published at Naples. Apparently, Stengel wanted to edify on the basis of factual information.[31]

In sum, Stengel's book refers to the Catholic tradition, but is clearly different from the university textbooks by his fellow Jesuits. The textbooks mostly ignore the supernatural elements, apart from the biblical earthquakes. Stengel on the other hand ignores natural causes, though without outrightly denying them. Here again one may wonder whether the written textbooks represent what was actually taught at the Jesuit colleges. In the cure of souls, earthquakes were definitely seen as divine punishments.

Catholics appear to have been less hesitant to attribute guilt to specific people than Protestants.[32] The instrumentalization of earthquakes in actual political or religious conflicts remains rare, but it did happen. An example is a French pamphlet from 1619 on an earthquake in the land of Béarn, at the foot of the Pyrenees. A serious disagreement had arisen between the French crown and the provincial *parlement*. Formally, Béarn was part of the kingdom of Navarre. Its earlier rulers had introduced Protestantism and expropriated the goods of the Catholic Church. Louis XIII, King of both France and Navarre, now wanted to return those goods, while the *parlement* resisted. A royal commissioner could not move them. The very day that the commissioner left, "an earthquake occurred, so terrible, that it seemed as if Béarn wanted to leave Béarn to follow him, in order not to be an accomplice to their disobedience. It occurred throughout the whole province and not beyond".[33] At the same time, a red cross was seen in the sky. The significance of all this hardly needed to be spelled out. The author, clearly a lawyer, warns that anybody who resists the King revolts against God, who may punish the perpetrators severely.

Whereas printed sermons are largely restricted to the Protestant world, major histories of earthquakes are mostly a Catholic affair. This is largely due, of course, to the fact that major quakes that would invite such histories are in practice restricted to southern Europe. Unlike the edifying treatises from northern Europe, such histories focus more on a detailed description of the event. The distinction is blurred, however, as under the influence of the Counter-Reformation, many histories contain religious and edifying elements as well. Already some of the writings on the earthquake of Ferrara contain such elements, and the tendency became more marked in the course of time. The humanist tradition was not completely repudiated, but it was definitely transformed. The dialogue form got out of fashion. Histories take the form of large descriptions. Among the major quakes that were discussed in this way are the earthquake of Apulia in 1627, the earthquake of Calabria in 1638, the earthquake of Sicily (Noto, Catania) in 1693, and the earthquake of Sicily (Palermo) in 1726.

Such histories would typically include a description of the quake, a list of afflicted towns and destroyed buildings, and notable events during the quake. Depending on the author's interests, they might also include a longer or shorter general discussion on the causes, effects, and so on of earthquakes, more or less on the model of the medieval commentaries and the Renaissance dialogues, and religious comments. Some authors managed to keep their work largely free from religious elements. Di Somma's history of the Calabrian earthquake of 1638 is rather written in a Stoic vein, without any reference to divine punishments or divine anger. Still, most of the large histories in the seventeenth century have a distinctly Counter-Reformation character. Many authors of such works were actually clerics. They give their works a religious character, not so much by adding all kind of religious comments, but rather by including in the description not just the damage to houses and the loss of life, but also miraculous rescues, the pious behaviour of the clergy and (as stated) the religious ceremonies that followed the devastation. Some episodes recur in several histories, like the person who was saved because a church bell fell over him or her, thus protecting the person from further harm.

Ruffo, in his description of the earthquake of Palermo, included many such stories. One, allegedly told by many people, was about the priest Giuseppe Galeoni. When after the first shocks, people ran out of their houses into the narrow streets, calling for confession and mercy, Galeoni went to his window and from there quickly gave the people the absolution. He had hardly finished when he fell down and with his pious flock was buried beneath the collapsing buildings. Another story is about a woman who was buried under the rubble and complained that she had actually deserved not just such a death, but hell itself. A priest took her confession and gave her the absolution before she expired. A concubine saw her lover next to her in bed crushed to death. She survived, sacrificed her hair at the chapel of St Nicholas of Tolentina, and started a better life. A woman fell on her knees at the first shock, implored God for mercy and invoked the Virgin Mary. The house collapsed over her, but it built a small cabin wherein she remained unharmed. Besides the behaviour of the clergy, these histories typically also praise the actions of the secular authorities to maintain order, clear the streets, and dig out the victims. Obviously, people were not free to criticize the actions of their authorities and such praise should probably be taken with a grain of salt.[34]

Like in the Protestant world, the reaction to prophecies and portents was ambivalent. Priests were happy to refer to portents that supported their interpretation, but wary of rumours that they did not control. Here again, the damage to church buildings and other prominent constructions drew much attention. In 1726, the tower of the Dom at Palermo was inclined "as if it wanted to pay its respects, without having collapsed, but still standing as a spectacle and memorial of the divine punishment or judgement".[35]

Notes

1 J.G. Taylor (1975) 244–245.
2 *Vollständige Kirchen- und Haus Music* (1611) 1024–1026, 1029. This was a popular book of carols that saw many editions between 1611 and the mid-eighteenth century.
3 K. Kircher (1594) Fi(v)–Fii.
4 Babst (1599) A3, A3v.
5 Hondorff and Sturmius (1597–1598) II, 349.
6 Osiander (1601) 5.
7 Aubéry (1585) 18.
8 Aubéry (1585) 36.

9 Examples: *Warhafftige unnd erbaermliche Zeittung* (1584); Twynne (1936) 37; Gross (1618) 14–17; Eckstorm (1620) 5–6; Wendeler (1691) 3–5; *Chronological and historical account* (1750), title page. Weinberg (1991) 71, note 11. Ehmer (1988) 190 states that the text was often quoted in sermons on the occasion of disasters. Schnurmann (2001) 256–259 regards similar reactions in 1692 as a rejection of the tradition that saw earthquakes as divine punishments, in my view incorrectly.

10 *Kurzer Bericht* (1591) 4; Volmarius (1591) par 12–13.

11 M.P.S.A.C. *Terra tremens* (1670) Mij.

12 Ashworth (1990).

13 Overvoorde (1907). The Bible was printed in Antwerp in 1534. Overvoorde notes that at his time, it was owned by a Catholic priest at Leiden.

14 *Historischer Kern* (1692) 134–135.

15 Baudartius (1624) book X, p. 103, book XI, p. 11. See also Eckstorm (1620) 11–15; Gross (1618), *passim*. For reports on Piuro, see Scaramellini (1988).

16 Ladero Quesada (1999) 100–101; Batlle (1999) 71, 73–74; Poirier (2018) 158. See also Guidoboni (1984) 115–122.

17 Udías (2009) 43.

18 Osorio (2008) 23.

19 Condorelli (2013) 151–152.

20 J.F.M.M., *Noticia* (1726) 7–8. Cf. also Rueda and Fernández (2008) 584–585.

21 *Noticia* (1733) 7.

22 Ruffo (1727, German version) 26–36 and Richter's preface to the reader. Other examples: *Relaçam nova* (1673) *passim*; Magnati (1688) 30–33, 241–249; Chracas (1704) 4–129; *Noticia* (1733) 7–8; *Relaçam* (1748) 5–7; Llano y Zapata (1748) 3–4, 8–10, 13, 19–21, 31–33.

23 *Discours espouventable* (1579).

24 Udías (2009) 43.

25 See Rueda and Fernández (2008) 585 for a sermon printed after the earthquake of Málaga. Condorelli (2013) 156–158, discussing sermons on the Sicilian earthquake of 1693, mentions in passing only one Catholic sermon.

26 Stalenus (1640) 16–20, 33. On Ephrem's treatise on earthquakes, see Dagron (1981) 89.

27 Stengel (1647).

28 Stengel (1651) II, 229–272.

29 Stengel (1651) II, 637–638. Stengel (1647) 22–24.

30 Stengel (1651) II, 232.

31 Stengel (1651) II, 237–238 (Constantinople), 241–244 (Jeruzalem), 244–259 (Calabria).

32 Condorelli (2013) 158.

33 *Declaration* (1619) 22.

34 Ruffo (1727, German version) 12, 13–14, 14–15, 17.

35 *Relation* (1726) 3 ("als wenn er eine Reverentz machen wolte").

11

MARGINALIZED APPROACHES

As argued, advocates of the confessional interpretation of earthquakes had to navigate between two extremes. They did not want to fall into the pit of naturalism, Aristotelian or otherwise, by which everything was explained from natural causes. On the other hand, they would not allow a completely spiritual interpretation of earthquakes, as they had to make sure that the interpretations were in line with their particular orthodoxy. On the whole, they operated quite successfully, but they were not able completely to eradicate competing views. It may illuminate the character of the confessional interpretation of nature to have a quick look at some of the views that remained on the margins.

Spiritual interpretations

The spiritual interpretation had been a powerful force in the early sixteenth century, with authors like Luther, Paracelsus, and Bodin, but the confessionalization process had reinstated scholasticism and philosophy. However, not everybody went along. The Hermetic and neo-Platonist tendencies, with their undogmatic, individual spirituality, were hard to accommodate with the growing confessionalization. Although increasingly marginalized, these remained a powerful source of inspiration for alternative, heterodox interpretations of the world.

The English physician Robert Fludd was among the philosophers who seriously tried to formulate an alternative for the dominant Aristotelian philosophy, largely along neo-Platonic lines. He did not ignore the visible world, but was inclined to interpret it in the light of his views of the spiritual world. Interestingly, Fludd too followed the general trend in paying much attention to meteorology. He wrote a "sacred and truly Christian philosophy, or cosmic meteorology", in four books. The work was published at Frankfurt (Germany) in 1626. The first two books deal with introductory material, general principles, and matter. The third book deals with the formal cause of the phenomena, and the fourth book with the efficient cause. The main theme of the work is the correspondence between the microcosm (man) and the macrocosm (the universe). Fludd's classification of meteors is complex and quite unlike any other author's. Earthquakes are classified as "dark, invisible, or occult meteors".[1]

Earthquakes are discussed both in book three and four. The chapter in book three, on formal causes, is titled: "On the earthquake and its accidents". Fludd starts by a violent criticism of the dominant Aristotelian view. He ridicules the idea that vapours can move the earth or would fight among each other. What could be the cause of such a thing? Maybe it is not strange that pagan philosophers, who were ignorant of God and His Word, maintained such opinions.

> But we assert that there is an aerial spirit in the earth, by which the whole terrestrial mass is alive, and is contained by it, in the same way as we see that the human body is preserved by its internal air; and this spirit can only be preserved if it continuously absorbs and takes in a little part of the world spirit.[2]

Fludd argued his case mainly by quoting Biblical sentences. From these, as well as from descriptions of earthquakes, he concluded:

> An earthquake happens when the *spiritus*, and as it were the soul of the total lump of the earth, confused by the regard of Jehova or the address of his angered angel, becomes incited and agitated so as to expand. It is then forced to break out violently from its terrestrial body into the open air, in the way of an enormous sigh, with a groan, or of ignited air from a cloud (…) and from this impulse is the terrestrial mass affected and moved, and the open air infected with evil and poisoned.

Immediately after that, he presents an alternative definition: An earthquake happens, when the terrestrial spirit, moved by Jehova's anger, "excited as if by fear and panic, makes its terrestrial body move and shake, in the way of the spirit of a man shaken by an unexpected fear".[3]

The material cause of the earthquake is an internal spirit. Fludd draws an analogy with the distillation of the spirit of nitre. "In the same way, the air enclosed in a dense cloud is animated and illuminated by the spirit or the mere seeing of the irate Jehova…".[4] From this type was the earthquake at Puozzoli (probably this refers to the Monte Nuovo) and the disaster at Piuro. These were accompanied by erupting flames, whirlwinds, lightning, and thunder. As for the quakes that have their cause in a non-ignited spirit, Fludd feels discharged from the obligation to discuss these, as the Bible talks about them at several places and many people still alive have witnessed them.

In the fourth book, on the efficient causes, Fludd makes a distinction between natural and supernatural (or spiritual) causes. Natural causes are either the elements or the heavens. Supernatural causes are either good or evil angels, or the first cause that ordains everything. Earthquakes are discussed both in the first section, on the first cause, and in the second, on good and evil angels. Chapter 21 is titled: "On the efficient cause of earthquakes, springs, storms at sea, and the like, and that this is only GOD, by the mediation of angelic organs." Fludd refers to Psalms 104:32 and Job 9:5 to argue that God is the primary and immediate cause of earthquakes. The Bible also tells (2 Sam 22:8) that earthquakes are brought about by God's anger, "for an earthquake is always working in the macrocosm *praeter naturam*".[5] God produces these quakes clothed in a cloud and surrounded by his angels, as is clear from many scriptural passages.

Fludd cites notably Matthew 28: 2–3, on the earthquake at the Resurrection:

> Suddenly there was a violent earthquake; an angel of the Lord descended from heaven; he came to the stone and rolled it away, and sat himself down on it. His face shone like lightning; his garments were white as snow.

According to Fludd, this text gives clear evidence that the angel of the Lord was the efficient cause of the earthquake. "The way that this happened was by his radiant appearance, that is, the brightness and essential splendor of God that shines forth from the angel's complexion." From this,

> it appears clearer than daylight that the opinion of the Pagans should be called impious and profane, that an earthquake is nothing else than the violent eruption of a multitude of winds contained within the bowels of the earth. Thereby they do not accept that wind and air are organs or vehicles of the divine aspect that thereby effects these stupendous things, and fall into a kind of great idolatry, attributing, in an injurious and vain way, the act of the Creator to the creature.[6]

That the traditional theories are idle could actually have been learned from the existing descriptions of earthquakes, both ancient and more recent. These tell about fire, plague, famine, and roaring sounds that accompany or follow earthquakes. It would be absurd to attribute these to some enclosed wind. Fludd seems to imply that such wonders are specifically "acts of God". He continues by discussing and refuting the views of the various philosophers more in detail. This does not tell us anything new.

In the end, Fludd is more interested in the acts of angels than in what we would call natural causes. Indeed, the second part of book four is devoted to the acts of angels. Fludd explains the differences between perverted and evil demons and elaborates on their hierarchy and order. His main point is that demons do harm only by the permission and the will of God. "God does not work by the mediation of angels, but by the Word in angels."[7] From the testimony of the Bible, he concludes

> that God is the sole author of the following meteors, to wit earthquakes, vapour or fog, lightning, (...), clouds, winds, rains, thunder, storms at sea,

etc., which he all in his wisdom produces by his angelic organs; but one should give special notice that as by the good angels he produces good things, and things that are useful and convenient for the earth, the waters, and the creatures, in a like way he uses his bad ministers, both in his anger against the impious, and for probing his elect.[8]

Fludd's statements about angels are in themselves perfectly orthodox. The unusual thing is his rigorous application of angelic theory to natural philosophy.

Fludd's work is interesting, as it demonstrates what kind of scholarship might have resulted if the spiritual and religious impulses of the early Reformation period had been given free rein. Although the religious interpretation is all-pervasive, that does not mean that physical considerations play no role. Where Fludd criticizes Aristotle, he does so frequently by invoking natural arguments.

Johan Baptist van Helmont, a physician from the southern (Spanish) Netherlands, in some respects is similar to Fludd. Both were physicians, working outside the university system. Both considered themselves as orthodox, Anglican in the one case, Catholic in the other, but had difficulty to be accepted within their respective Churches. They both continued the spiritual, Hermetic bend of the Reformation. But whereas Fludd regarded the world in the light of his esoteric knowledge and wrote books of cosmological speculation, Van Helmont was much more interested in empirical reality. He conducted many experiments and remains famous for his contributions to chemistry.

Van Helmont wrote a treatise, *Terrae tremor*, at the occasion of an earthquake that was felt in the Low Countries on the morning of April 4, 1640, when he was at Malines (or Mechelen). It was published posthumously in 1642 by the author's son, Franciscus Mercurius van Helmont, in *Ortus medicinae* [Rising of medicine], a compilation of various treatises by Van Helmont sr.[9] He wrote the treatise, he explains, out of dissatisfaction with the common explanation of the proximate (secondary) causes of the phenomenon. (On the primary cause, God, of course no disagreement existed.) As he said, all modern authors simply follow the theories of Aristotle. "If these would appear to me to be in agreement with the goals of the Divine Godhead, I would have abstained from writing." However, Aristotle's theories appear like figments. Van Helmont's first goal therefore is to demonstrate the impossibility of Aristotle's theory. His second, to explain his own opinion, "not founded upon pagan dreams, but confirmed by the doctrine of a higher authority".[10]

In the following paragraphs, Van Helmont gives a detailed and devastating criticism of the theory of exhalations. His criticism is based on the state of the art of seventeenth-century science. That there is not enough space for those exhalations, is argued with what we know about the constitution of the earth. That exhalations cannot be excited the way Aristotle teaches, is shown by seventeenth-century chemistry. Explosions seek a breakthrough at the weakest spot, they do not lift a whole country. Since the invention of the thermometer, it is known that caves have a constant temperature throughout the year. "So, I have no doubt to deny the natural cause given by the schools, and invented by the devil to obfuscate my God's honour."[11] He ridiculed the ignorance of the scholastics, denied the influence of the stars, and emphasized the difference between volcanic eruptions and earthquakes.

Van Helmont emphasized that earthquakes were portents. He referred to a manuscript of the curate of the Church of St. Mary at Malines, who pointed out that in 1540, the earth had trembled for three consecutive days before the powder house at Malines had exploded; and that there had been an earthquake in 1580, two days before English troops had conquered the city (the so-called "English Fury" in April 1580). "But what do these events, happening from fatal necessity, in the chain of causes have in common with that dreamed-of subterranean exhalation?" Exhalations do not have such a meaning. Likewise, the eruption of Mount Vesuvius in 1631 was announced by an earthquake. The eruption itself had natural causes. But these preceding signs had nothing in common with the fire. Van Helmont's conclusion is that the earth was not moved by something like air, but rather in the way a bell was struck by the clapper. "Therefore, I believe that the earth trembles and is frightened, every time that the Angel of the Lord strikes it." Like Fludd, he referred to Matthew 28.[12]

Van Helmont continued by producing some arguments for this assertion. First, immaterial spirits work by divine powers, so that nothing can resist them. Matter is powerless against the spirit. Still, just like the angel of the Apocalypse uses a bowl to pour out extermination, so he uses a sound to produce an earthquake. Van Helmont refers to the sound that accompanied the earthquake at Malines. He sees an analogy in what happens during a thunderstorm: lightning burns and melts, but it is the thunder that strikes. This is argued partly from the Bible, partly from common lore about the effects of thunder: silkworms die, milk and beer are spoiled, etc. "So, let the voice of the thunder and the voice of the earthquake be the sounds of administering spirits."[13]

Other accompanying effects of earthquakes are mere accidents. Against the objection that some regions are more liable to earthquakes than others he asserts that one cannot draw conclusions from this, as these things are subject to the will of God. That after an earthquake sometimes noxious vapours rise from the earth happens naturally and accidentally and is not related to the cause of the earthquake.

> This blindness for the causes of earthquakes, by which mortals lay off all fear of the Godhead, has been invented by the devil, so that they will invert the goals that God has set them (to wit, piously honour His majesty and bring back to mind the crimes of the life they have led), if not completely forget them.[14]

Van Helmont also rejected the story that someone had predicted the 1640 earthquake from the smell of sulphur that the three preceding days allegedly had risen from a very deep well in the castle of Louvain. He objected that the well looked deep because the castle stood very high, but that actually it did not penetrate deep into the earth at all, not to any depth where sulphur could be expected. So, this event simply could not have happened. Of course, God might have given a sign, but this does not indicate a natural cause of an earthquake throughout the Low Countries.[15]

Van Helmont concluded with some religious considerations. The earth trembles because it has been struck by God. The goal of an earthquake is that the sinner will convert. Van Helmont made a long digression on the various punishments that belong to different sins. Earthquakes happen especially because of the sins of the blood calling out to heaven for vengeance. They therefore announce punishments appropriate for mercilessness, cruelty and injustice. This is something good, although springing from a bad cause, as it gives the impenitent a warning. The earthquake after Christ's resurrection announced the destruction of Jerusalem and the Jewish nation.[16]

In a final section, Van Helmont replied to some criticisms by a friend to whom he had given the unfinished treatise for reading. This friend allegedly admitted that the earthquake happened supernaturally, but felt it was not justified to explain the very way God had produced it (an angel striking the earth by a sound, according to Van Helmont). God has a thousand ways to effect things. Why could he not have used wind or gunpowder, or struck the earth himself? Van Helmont answered that God, when acting in a miraculous way, might use natural means, but not so that these means could seem supplementary causes. That wind or gunpowder played a role was impossible in nature, as he had shown. Moreover, he had not suggested his causes from mere speculation, but from Scripture: "The voice of the thunder lashed the earth" (Sirach 43:18), and other texts.[17] He specified that he did not publish his treatise in order to show himself clever, but in order that the trembling of the earth might awake in us the fear of the Lord.

All in all, Van Helmont's treatise is hard to categorize. He rejects any naturalistic interpretation, but the arguments are mostly taken from empirical science.

Scepticism, libertinism, stoicism

Not everybody accepted the ecclesiastical interpretation at face value. Not only there lingered all kinds of older traditions among the population at large, among the learned too there was some degree of dissatisfaction with official orthodoxy. In many cases, the problem was not so much the intellectual content itself, but rather the political and ecclesiastical order that these ideas embodied. Many people felt that the confessional ideas mainly served the interests of the clergy. Since the official view was propagated by the powers of Church and state, people could rarely afford to venture their objections openly. However, there were indirect ways of showing one's doubts.

Instead of openly disagreeing with the views propagated by the Church, one could draw attention to relevant passages in ancient authors, as these were well respected and could not easily be dismissed. Criticism on the belief that comets were portents was put forward with reference to ancient authors. For earthquakes, one could especially refer to the Stoic tradition and the works of Seneca. Seneca was a well-respected author and ecclesiastical authors could use him for edifying purposes.[18] However, not everybody was impressed by his piety. The minister Walter Cross in 1692 felt that what Seneca's said about earthquakes was "a great Argument against *Seneca*'s Christianity, and true Piety too".[19] Another author who was sometimes referred to in this respect was the second-century Roman author Aulus Gellius. In his *Noctes Atticae* (book II, chapter 28), he had stated that the cause of earthquakes is uncertain and that it is even unclear to what deities one should sacrifice in the case of an earthquake. This text could be used to express a general scepticism on the causes of earthquakes.

Among the counter current are the French *libertins*, among whom François de La Mothe Le Vayer, a French magistrate, takes a prominent place. He wrote many philosophical and historical works. Though not openly denying the theological tenets, he wrote about them in a rather sceptical vein. Among his writings is a *Physique du prince*, which also contains a chapter on meteorology. He dismissed the prodigious character of presumed blood rains, referring to Wendelen and Gassendi, and stated that the rainbow definitely must have existed before the Deluge. He subtly mocked the Conimbricenses ("whom I otherwise esteem very highly") for their view that the wind that ladies produce with their fans is no real wind. As to earthquakes, La Mothe Le Vayer simply refers to another work that he has written on the topic.[20]

This treatise, *Des tremblemens de terre* [On earthquakes], is in the form of a letter, written on the occasion of the destruction of the Alpine town of Piuro (or Plurs) in 1618, an event that caused a lot of attention throughout Europe.[21] It is rather short: not even four pages. Unlike similar treatises, La Mothe Le Vayer did not rehearse the various theories on earthquakes. Rather, he stressed our ignorance. He praised the ancient Romans, who, as related by Aulus Gellius, sacrificed to unknown gods and thereby acknowledged their ignorance. On the other hand, he mocked the ancient Greeks who claimed a superior knowledge and gave rules for predicting earthquakes: "Is it not that considering the earth as a big animal, they possessed the art of feeling its pulse and in that way knowing the convulsions that were to happen to it?"[22]

The next section is more traditional. Here, La Mothe Le Vayer passed in review some examples that were particularly strange or otherwise noteworthy. He denied that there were territories where earthquakes did not occur: in claiming this, the nation of philosophers was as credible as the poets, he stated. All countries were subject to earthquakes, although the very hot and the very cold somewhat less so. When he came to talk of the effects of earthquakes, he was again rather unconventional, as he came up with some examples that earthquakes had worked favourably for some people. Double-edged is his remark that these extraordinary concussions would be rather welcome, if they only hit the most wicked and let the good people unharmed. As for the view of earthquakes as bad omens, his scepticism was unmitigated. One general drew his army back after an earthquake, another saw an argument to continue the war. "This shows the instability of the human mind concerning the instability of the earth, as concerning all other things."[23]

La Mothe Le Vayer concluded that we should not be too shocked by an earthquake. If we accept that the earth is moving in 24 hours (not talking about the other movements), why would it not shake as well? It would be strange if it would not. So, he did refer to new scientific discoveries, the theory of Copernicus in this case. It should be noted, however, that the Copernican theory was still controversial at this time. The reference rather served the purpose of hinting at the uncertainties in our knowledge, than appealing to science as a solid alternative to a theological view. As for the dangers of an earthquake, he reiterated the argument of Seneca: if we are to die, why not from something big? Petty causes can bring death in a much more painful way. Finally, he gave the practical advice (referring to pope Boniface VIII) that in case of emergency, one could build a light wooden construction, which would cause no danger when collapsing.

Interestingly, sceptical views were not confined to libertine circles outside the university. Some professors and students seem to have entertained similar ideas. Open defiance of the orthodox tenets was of course out of the question, but people knew how to read between the lines. One could give expression to one's reservations in a way that was formally impeccable, but that would be understood by one's friends and sympathizers.

At Leipzig in 1648, the professor of philosophy Christian Friedrich Franckenstein defended a physical disputation on earthquakes in the form of some observations on book II, chapter 28 of the *Noctes Atticae* by Aulus Gellius – as said, a somewhat sceptical text. In the introduction, Franckenstein, referring to Seneca, explained that his intention was to take away the fear of earthquakes by investigating their causes. He also included Seneca's observation that death will befall us anyway. Franckenstein referred to Pythagoras who according to some sources had stated that earthquakes were caused by the spirits of the dead, but stated that this opinion would not be worth discussing, had Bodin not revived it. But Cicero rightly stated in his work on divination: "whatever comes into existence, of whatever kind, must needs find its cause in nature".[24] For the rest, Franckenstein follows the traditional theory of an exhalation generated by the subterranean fire.

The disputation nearly exclusively focusses on the efficient cause of earthquakes and omits any religious dimension. Franckenstein announced that he would write a sequel on the significance of earthquakes, but as far as can be ascertained, this never saw the light. At the end of the disputation, some additional theses were added (as was common). Here, it was added that there were supernatural earthquakes as well, as those at the crucifixion and the resurrection. On the other hand, on prodigies that predicted impending evils, the author said that the more these were believed, the more they were imagined to happen.[25]

At Jena, a certain Philipp Georg Luck wrote and defended a disputation under the professor of philosophy Johannes Paulus Hebenstreit on the occasion of the great Sicilian earthquake of 1693. This disputation is nearly exclusively on physical aspects. Luck discusses the nature of Sicily, the physical causes of earth-quakes, and why Sicily is so liable to them. At the end, however, he announced a sequel on the effects and specifics of the earthquake wherein he also would ex-plain the anger of God in these events, in order not to remain stuck in mere nat-ural causes. Finally, he rejected Seneca's view that earthquakes were not sent by the gods, but have causes of their own. Luck's disclaimer sounds a bit obligatory. As far as is known, as in the case of Franckenstein's disputation, the promised sequel never appeared. Strikingly, Seneca's opinion, with which he purportedly disagrees, concluded the disputation, giving the Stoic the last word.[26]

Whatever the exact views of these authors, they did publish ideas on earth-quakes that deviated from the standard Lutheran discourse. Intellectual life at Protestant German universities was clearly richer than the domination of Lu-theran orthodoxy would suggest.

Popular naturalism

One may well ask whether the message that earthquakes are portentous and acts of God, reflects a widely held, unopposed belief, or whether preachers insisted so much on this point because they knew it was actually contentious. There are indications that actually the latter was the case. The ministers themselves recog-nized sometimes that their views were met with opposition, obviously by people they dismissed as impious or even atheists.

As we saw earlier, David Schweizer in his sermon of the earthquake of Vienna in 1590 did denounce people who felt that earthquakes were just natural things that did not announce any harm. He definitely referred not just to classical opin-ions, but to views that purportedly were common in his own time. Likewise, the minister Johannes Moltherus in a sermon in 1601 complained that some people feel that earthquakes are mere natural events:

> One can find even today such people, who state that the earth is of a kind that it purifies itself in this way. As one can see in trees and in the earth itself, that they purify and clean themselves from superfluous matter, with resin, mushrooms, moss, and the like.[27]

This view is so outlandish that it is hard to imagine that Moltherus just made it up. A treatise from 1670 likewise states that there were people who pretended "such shaking of the earth (and also comets), were natural, therefore there was no reason to be afraid".[28]

Such voices remain anonymous, but some opinions were described in a little more detail. Schweizer, in denouncing his opponents, specified that these people "have heard from their ancestors that similar earthquakes have occurred years ago, and that these were followed by good years, fertile in crops and cattle".[29] One might surmise that Schweitzer, in order to strengthen his point that earth-quakes were acts of God and forebodings of evil, was attacking a straw man, but actually, the idea that earthquakes announced good fertile years is mentioned by other sources as well. The first mention is in a newsletter of the banking house of Fugger, on the occasion of a small earthquake at Vienna in June 1590: "Several old people kept it for a presage of a good, cheap, fertile year, followed by a high mortality (as also happened several years ago)."[30] On the occasion of the big earthquake of 15 September, not just Schweitzer denounced this opinion. The ministers Conrad Kircher and Marcus Volmarius also mentioned it. Volmarius even claims it as a main reason for publishing his pamphlet.[31] Another mention is made in a sermon by the minister Tobias Wagner on the occasion of a series of earthquakes in Tübingen, in southern Germany, in 1655.[32]

Whereas these ministers refuted this view as godless, others reported on it in a more neutral manner. Bartholomaeus Keckermann in his disputation of the earthquake of 1601 made passing reference to it, as a strange but apparently harmless opinion.[33] It is also mentioned in the learned journal *Miscellanea Curiosa*, in a report from Ljubjana on an earthquake in Slovenia on 19 February 1691, where it is equally mentioned among the popular opinions that were ventured.[34] There seems enough documentation to establish that the idea that earthquakes announced fertile years was a current opinion in the Alpine region, probably foremost among the peasant population. This represents an oral tradition that ran counter to the established (counter-) reformation views.

Whereas the above-mentioned ministers, although regarding the view as godless, paid only scant attention to it, there happened to be one author who refuted it at greater length. The physician Johann Burgower (or Burgauer) from the city of Schaffhausen on the upper Rhine wrote a treatise on the 1601 earthquake. Although a physician, not a minister, Burgower's interests appear to have been mainly religious. He is known to have written several purely theological treatises, which however remained unpublished. The treatise on the earthquake was only published by his heirs in 1651. The tenor of this work too is religious. He started in a conventional way by listing current physical explanations, but then simply rejected them on the grounds that the Bible, when speaking of earthquakes, does not mention such exhalations as mentioned by those theories. The Bible "throws down all above mentioned causes of the philosophers".[35] His true target however are not so much the opinions of ancient philosophers, but the idea that earthquakes are good omens, announcing fertile years. Refuting this idea and stating that earthquakes are unambiguously bad signs appears to have been his main motive in writing the book.

Unfortunately, Burgower does not identify his opponents, but the information he offers on their ideas is quite detailed, which makes it likely that he did have some specific people in mind. Allegedly, the opinion that earthquakes are good omens was entertained not just by "the common mob", but also by people "who keep themselves for better and wiser than the common populace". They feel, according to Burgower, that their view "is not a mere speculation, but founded upon experience and history, yes, on the very Scriptures".[36] After all, the earthquakes mentioned in the New Testament meant no harm, but good:

the earthquake at the Crucifixion converted the Roman centurion, Paul and Silas were liberated from their dungeon by an earthquake, and an earthquake preceded Christ's resurrection. As for history, the supporters of this view refer to the chronicle by Johann Stumpff to demonstrate that the earthquake of Konstanz at 1277 and the earthquakes of Basel in 1444 and 1533 were all followed by cheap and fertile years.[37] Based on what Burgower writes on his opponents, it would seem that these people were certainly no scholars, but still literate, having access to the Bible and to vernacular works of history. They seem to have been some forerunners of the lay philosophers whom we will meet in a later section.

It is hard to determine how successful the campaigns of the Reformation and Counter-Reformation were to propagate the orthodox view of nature. Many people indeed appear to have accepted the orthodox view. The Utrecht lawyer Aernout van Buchell after an earthquake in 1640 noted in his diary that in its wake all kinds of portents were reported. These were apparently spontaneous sightings by the people.[38] Still, it would have been strange if nobody had privately denied the prodigious nature of earthquakes. Obviously, dissenting views could not be printed. The fact that nevertheless there appear some common views indicates that the confessional views of nature was not a self-evident, age-old view, but that it was actively imposed by the proponents of the Reformation and Counter-Reformation in the sixteenth century. In this way, they created a new science, distinct from the medieval one, and a new popular understanding of nature.

All this happened long before the rise of a mechanistic view of nature. The philosophical explanations that were current in the Middle Ages did refer to natural causes, but did not leave God or Providence out of the game. Nature was hardly seen as an independent entity. Still, the view that specific phenomena should be attributed to "nature" was refuted over and over again in the Reformation era. One wonders whom these attacks were exactly directed against. This anti-naturalistic streak would dominate the discourse on natural phenomena by both Reformation and Counter-Reformation authors for a long time to come. By constantly pitting supernatural and natural explanations against each other, the preachers may well have given the impression that God mainly resides in miracles and has little to do with the ordinary course of nature. This may have led to a growing awareness of the potency of natural explanations.

It seems certain, anyway, that the (Counter-)Reformation offensive resulted in a transformation of the medieval view of nature. By interpreting strange events as direct intrusions of the divine into the ordinary world, these were placed outside the normal order of things. Strange events were no longer seen in relation to other human affairs, in a symbolic, allegorical or exemplary way. They were witnesses of God's transcendent power. As such, they were used to legitimate the doctrine of the Church and the moral order of society, but Church leaders were reluctant to give them more specific meanings. Removing them from the world of ordinary human affairs would eventually undermine the idea of the world as a web of correspondences, whereby all elements and events had some meaning beyond themselves.

Notes

1 Fludd (1626) 104. Another, more traditional scheme of meteors on pp. 145–146 does not include earthquakes.
2 Fludd (1626) 131. The chapter on formal causes is on pp. 131–132.
3 Fludd (1626) 132.
4 Fludd (1626) 132. Chapter 21 is on pp. 204–207.
5 Fludd (1626) 204.
6 Fludd (1626) 204.
7 Fludd (1626) 246.
8 Fludd (1626) 247.
9 van Helmont (1648) 92–103. I also used the German translation by Christian Knorr von Rosenroth: van Helmont (1683/1971) 131–142.
10 van Helmont (1648) sect. 4.
11 van Helmont (1648) 98, sect. 23.
12 van Helmont (1648) 100, sect. 28, 30.
13 van Helmont (1648) 101, sect. 30.
14 van Helmont (1648) 101, sect. 30.
15 van Helmont (1648) sect. 31.
16 van Helmont (1648) sect. 32–36.
17 van Helmont (1648) 103 (sect. 37). The text as quoted by van Helmont: "Vox tonitrui verberavit terram". The new translation reads (Jezus Sirach 43: 16–17): "The crash of his thunder makes the earth writhe, and, when he appears, an earthquake shakes the hills (…)".
18 See e.g. Stalenus (1640) preface, 16. Grimaldi (1703) 35–38.
19 Cross (1692) 17.
20 La Mothe Le Vayer (1662) I, 978. The full treatise is on pp. 933–1015.
21 La Mothe Le Vayer (1662) II, 715–718.
22 La Mothe Le Vayer (1662) II, 716.
23 La Mothe Le Vayer (1662) II, 717.
24 Cicero (1971) 439 (*De divinatione,* book II, chapter 28, par. 60).
25 Disputation Leipzig (1648 Jan. 25).
26 Disputation Jena (1693 Febr.).
27 Moltherus (1601) 8.
28 M.P.S.A.C, *Terra tremens* (1670) preface.
29 Schweitzer (1593) 33; also printed in Gutdeutsch et al. (1987) 145. Schweitzer returns to the point on pp. 77–78.
30 Printed in Gutdeutsch et al. (1987) 115–116.
31 Kircher (1594) [Biiij]v, Cv, [Hiiij]v. Volmarius (1591), preface and par. 1.
32 Wagner (1655) 62. Wagner's sermon, according to his own testimony, was largely based on an earlier sermon, from 1601, by his predecessor Sigwart, which I have not retrieved, but which is discussed by Ehmer (1988) 191–192; see also 196, for Wagner's sermon. Ehmer does not mention the presumed impact of earthquakes on the harvest.
33 Keckerman (1607) 65. See also Rouffignac (1694) 21.
34 Thalnitscher (1691) 425–427.
35 Burgower (1651) 175.
36 Burgower (1651) 205.
37 Burgower (1651) 205–208. Johann Stumpff (1500–ca.1574) published in 1554 *Schwytzer chronica,* an abbreviated version of his unpublished *Schweizer- und Reformationschronik.*
38 van Buchell (1940) 107–108: "Variae ostenta feruntur post terraemotum nuperum, nubium in caelo pugnam Arnemiae et quae non alia prodigia…". The earthquake itself is mentioned on p. 105; at that place, van Buchell only mentioned that earthquakes are uncommon in his region.

PART III
The rise of modern empiricism

12

NEW SOURCES OF INFORMATION AND THE RISE OF A SCIENTIFIC PUBLIC

The confessionalized view of the world had been extremely successful, but by the end of the seventeenth century, it got into serious difficulties. It still had its defenders and dominated a large part of popular culture, but its detractors made themselves heard ever more loudly. The view that comets or monsters should be regarded as omens that announced future disasters was openly criticized. A new view of nature imposed itself at universities and in learned circles, which banned final causes and specific meanings and taught that all of nature was ruled by immutable, universal laws. This new philosophy got strong support from the impressive results that investigators of nature now obtained while applying these rules.

In traditional historiography, the demise of older views of the world was seen as a direct result of the new inventions and discoveries made by seventeenth-century scientists. No doubt there is some connection there, but it is far too simple to see the one as the result of the other. For instance, more recent research suggests that the refutation of the prodigious nature of comets owed more to traditional humanist scholarship than to modern scientific insights. The changing view of monsters in the eighteenth century has been connected to social rather than scientific trends. The new view of nature resulted from new religious sensibilities as much as from new scientific discoveries.[1]

Confessionalized philosophy had been very much a product of the political and religious situation at the beginning of the sixteenth century in Europe. In order to understand its decline, we likewise have to look at this wider context. Confessionalized philosophy relied on a situation wherein adherence to a specific Church doctrine had become essential for one's political and social identity. Church life thereby impregnated all intellectual activities. However, the consensus which had supported the symbiosis between the Church and the state was gradually eroding. States, growing more powerful, increasingly resented their dependence on the local Churches. This encouraged a more critical view of the ecclesiastical order and its underlying ideology. The confessional philosophies would not disappear, but they would increasingly be faced with alternative views of the world. In particular, the period 1680–1715, the "crisis of the European mind" as Paul Hazard called it, proved a watershed. Critics of orthodoxy had never been completely absent, but earlier they had been pushed to the margins. As uneasiness with the dominant role of the Church in the state grew, alternative views again got more attention. Within the Churches, new interpretations came to the fore.

In this last part, we focus on the views of earthquakes in the period of confessionalized philosophy's decline. In the first chapters, we will have a look at the changing intellectual landscape and the various engines of change. The focus will be on the influence of new ways of communication and the availability of new information. Early confessionalized philosophy had been successful because of its control of the printing press. A hundred years later, the sources of information had considerably expanded and this obviously left its mark not just on general knowledge, but also on the explanatory frameworks available. New media propagated a "new empiricism", which not only directly influenced how scholars studied and discussed earthquakes, but also was an important precondition for the new sciences in general. That is not to say that philosophical movements were unimportant, as they legitimized the new practices. But focussing on information rather than theory hopefully will describe the interaction between the new and the old worldviews in a somewhat more sophisticated way than as a simple opposition. After this, in the next chapters, I will investigate how confessionalized philosophy fared in this age of new information and upcoming sciences, both in the case of scholarly views and the more popular use of them.

A widening world

People were understandably most concerned about earthquakes that they had witnessed themselves or that occurred in their own land. There was also, especially in Germany, interest in news from the Near East. We mentioned above the reports on the earthquake of Constantinople of 1509. A pamphlet from 1542 on the earthquake of Scarperia in Italy includes a short message of an earthquake in Turkey. The year 1546 saw the publication of a pamphlet at Wittenberg on an earthquake in Palestine and a tempest in Famagusta on Cyprus. There are also pamphlets on an earthquake with showers of fire and thunderstorms in Constantinople, and on an earthquake and rainstorm in Mecca in Arabia.[2] These concern places that were important on the mental map of sixteenth-century Germans, as it was shaped by Biblical and ecclesiastical history and the wars with the Turks. The Earthquakes in Muslim countries could be seen as signs that God was on the side of the Christians. The pamphlets demonstrate God's hand in history rather than that they give a description of natural or geographical facts.

However, when in course of time European travellers explored the overseas world, more information did come in that did not necessarily fit well within the biblical framework. Reports of the overseas wonders were much in demand, and these included overseas earthquakes. Accounts of such events, however, would be different from reports of things closer to home. As argued before, reports of European earthquakes generally had a description of the event itself, focussing on what happened to important buildings or prominent persons, put within an explanatory theological framework. The theological explanation was important to make sense of these events. Disasters in the overseas world did not fit well into the established narrative. There was less empathy with the local population and the local monuments that got damaged were of little appeal to the European public. Moreover, the story of sin, punishment, and repentance appeared not well to apply to a population who were pagans anyway.

European witnesses would feel themselves to a large extent outsiders, which alone would cause a more detached view. (Of course, this was not necessarily true when it concerned European settlements – the various reports on earthquakes in Peru, or the reports on the earthquake in the English colony of Jamaica in 1692 follow very much the standard narrative.) Descriptions of overseas countries were written for entertainment purposes, not for moral instruction. Relations of earthquakes or similar events would rather fall within the category of travel reports, which normally were printed without an edifying introduction or conclusion.

Moreover, learned clergy, who played such an important role in the interpretation of earthquakes in Europe, were largely absent overseas. The people most directly affected by events there, and most likely to report on them, were seafarers and merchants, for whom such events in the first place presented a business risk. They could not afford to look upon them with awe or devotion, but had to calculate their losses and opportunities. Harold Cook has drawn attention to the way Dutch merchants who went overseas were attracted to "objective" knowledge. The case of earthquakes seems to confirm his thesis. Reports on earthquakes or volcanic eruptions from the colonies are mostly concerned with the economic consequences of the event and often completely silent about the religious aspects.[3]

Early modern travel literature is far too extensive to discuss here in detail, but some cases might illustrate the mindset of the authors and the message their readers might get. In March 1615, the fleet of the Dutch United East India Company arrived at the island of Banda Nera, where they had earlier established an outpost. As it happened, this was just days after a major eruption of the volcano on the island. In a report to the directors of the company back home, the governor, Van Reynst, described what he found:

> I have found the castle Nassau (…) outwardly in perfect order. However, the roofs of the houses had most collapsed by the eruption of the Gunung Api or Sulphur Mountain, which is nearby. On the sixteenth, the mountain has bursted with very violent blows, throwing out big glowing stones and much sulphurous dust. Part of the mountain has been turned upside

down, so that the ships on the roadstead seemed about to be torn asunder. The trees of Nera have been completely overblown by the smoke, so apparently there wil be no nutmeg for the next year or two. It was a very strange sight. The water stood all around the mountain and was boiling and the fish was shovelled by baskets full, mostly dead. The people from [the nearby island of] Lontor felt that we should have abandoned the castle, as indeed we should if the smoke had continued.[4]

Clearly, Van Reynst was not concerned with the theological significance of the eruption, but with its effects on business.

The same practical approach to similar events transpires in printed reports for the general public. So, a Dutch pamphlet relates, in the form of a letter, a voyage in 1652–1653 to several islands in the Atlantic, among others to the Island of Sao Miguel in the Azores to take in grain. The anonymous author appears to have been an Amsterdam merchant, but the ship was sailing under the flag of the southern (Spanish) Netherlands. The Dutch at this time were at war with the Portuguese, who ruled the Azores, as well as with England. English men of war were hunting for Dutch merchantmen, so it made sense to go undercover. The author clearly felt a stranger in the Azores.

After arrival at Sao Miguel, the author went ashore to manage the loading of the cargo, that is, measure the grain and make the payments. That very day (October 13, 1652), the earth started to shake. Locals told that some lighter tremors had been felt the day before. The earthquakes continued the next days, getting worse and worse. The people of the island were afraid, many slept in the fields. Processions and penitences were held. As the author writes, everybody prepared to die. "One heard nothing but crying and weeping, friends and enemies begging each other forgiveness." It is striking that he says nothing about his own feelings, nor does he spill ink on any pious thoughts. Rather, he complains that under the circumstances, as everybody was praying and crying, it was very hard to get labourers.[5]

As it turned out, the quakes formed the introduction to a volcanic eruption on the island. On October 19, news came that the earth had opened, about one and a half mile from the city. "The dejection [*verslagenheid*] that this caused among the people cannot be told, let alone written … for they firmly believed that the whole island would have sprung asunder that night." The author thereupon, together with the ship's captain, rented horses to take a look at the place of the eruption. At the crater, they investigated the stones that had been thrown out, in the hope of finding some precious metals, "but everything had been completely consumed by the fire and was worthless".[6]

Just like events in Europe were compiled into large chronicles, the interest in the overseas world soon led not just to the publication of individual travel reports, but also to elaborate descriptions and histories wherein much information was compiled. These became standard repositories of European knowledge of the overseas world. Following their sources, these compilations described earthquakes or other disastrous events in a rather factual way, without the usual theological interpretations. The natural aspects in the description got the upper hand over the providential ones.

A prolific author of travel reports and geographical descriptions (all works of compilation) was the Dutch minister Arnoldus Montanus. In 1669, he published a history of the diplomatic missions which the Dutch, who had a trading post at the Japanese island of Deshima, made every year to the court of the Emperor. This was one of the first descriptions of Japan available in Europe. Japan is a very earthquake-prone country and Montanus did not fail to include them in his work. Among other things, he included a description of a major earthquake which had recently struck the Japanese city of Jedo (Edo, present-day Tokyo). The topic figured rather prominently in the book, as it was accompanied by a large print on two pages. Apart from the description itself, Montanus included some theoretical considerations. He also briefly mentioned the popular beliefs of the Japanese on such earthquakes, which he of course regarded as superstition.[7]

Montanus was a hack writer who often repeated himself. He preferred interesting anecdotes over solid facts or theologically sound discussions. Still, even such pedestrian works would contribute to people's view of the outside world. The more such descriptions were published, the more people became aware that earthquakes and other unusual phenomena on a global scale were actually pretty common events. Descriptions of European earthquakes traditionally focussed on human vicissitudes and experiences. In the case of far-away countries, it was harder to collect such stories whereas readers would have more difficulty to identify with those foreign people. Reports on earthquake in far-away countries therefore had to focus more on the physical point of view.

New forms of publication, new means of information

The widening view of the world went hand in hand with the emergence of new forms of publication. Sixteenth- and seventeenth-century pamphlets addressed themselves to a general public and were heavily censored by the ecclesiastical and secular authorities. Merchants and public officials, however, increasingly felt that they needed more prompt and factual information on events that could affect prices, the safety of travel routes, and so on. By the end of the seventeenth century, there emerged a new print medium to serve their ends: the periodic newspaper. Newspapers did not print important items separately with a commentary, but collected and released the bare facts. Accuracy was more important than interpretation and religious instruction was no longer a main goal. Implicitly, they created an image of the world consisting of useful facts, rather than moral and biblical values.[8]

Most of the items in these newspapers concerned war and politics, but natural disasters were also included, as these too could affect business. As a matter of fact, such accounts initially were very short and sober, often only one or two sentences. A rather early, somewhat longer example can be found in the *Amsterdamse Courant* of 8 June 1688:

> About the earthquake at Lima in Peru, one has some more information from letters from London. It should have happened 22 October of last year. The city, besides two smaller towns, one of which apparently must be Callao di Lima, the harbour of Lima, are said to have been destroyed by a big earthquake and flood. Ships have drifted up to three miles inland by the tidal wave, and many corpses have washed ashore, so that incredible damage must have been done. Some, however, doubt the reports, although they have been confirmed by the English court, because they have not been confirmed from [the Dutch colony of] Curaçao. Moreover, there are letters from January 8 of this year 1688 from Porto Velo and Panama that do not mention it. Whether we can get any confirmation from this ship lately arrived in Zeeland [mentioned earlier in the paper], we shall have to learn from the next letters from Zeeland.[9]

The regular appearance of newspapers greatly enhanced the stream of information. David Wendeler in 1691 could refer to what had happened during the earthquake of November 1690 in many places in Germany as known from newspapers.[10] Newspapers could get their information directly from letters, but quite often they copied each other's reports. On 5 January 1699, an earthquake happened in the East Indies. An account was printed in Batavia (present day Jakarta), the Dutch colonial capital on Java, and sent to Holland, where an extract was printed in October in the *Haarlemsche Courant*. From Holland, the news was brought to England and an "Epitomy" was printed in English weeklies. Robert Hooke in his lecture for the Royal Society of January 10, 1700, gave the full account as it had appeared in the *London Post* of September 30 [1699]. The item is dated Amsterdam, October 2 (mind that Holland and England still followed different calendars) and refers to letters from Batavia dated February 8.[11]

Talking about the rise of newspapers as a clear-cut development is somewhat misleading. The publication landscape of the seventeenth century was extremely messy, with all kinds of periodicals competing for the attention of the public. Germany since the late sixteenth century had its *Messrelationen*, published on the occasion of the Frankfurt and Leipzig book fairs, which summarized political events since the last fair. An author who in 1670 referred to the landslide at Salzburg on July 16 the preceding year, noted that there had been printed several accounts, "but because these maybe went a bit too far in the first rush, before one was able to have the true details", he included the description as now given by the fall relation of the Frankfurt book fair.[12] As we saw before, the Dutch minister Baudartius also partly relied on information from the *Messrelationen*.

Another genre that emerged is the so-called *Mercurius*, basically a periodical that summarizes the news of the preceding period. They ranged from small ephemeral leaflets to real journals. Especially the latter allowed for a somewhat more elaborate discussion of events. The journal *Europische Mercurius*, published at Amsterdam, in 1692 reported on the big earthquake of Jamaica, then an English colony, in June 1692. It printed two letters from the Anglican minister at Port Royal, who described the event as an eyewitness. These letters had earlier been published in London. Later in 1692, the *Europische Mercurius* reported on the earthquake of 18 September that had been felt in Holland and neighbouring countries. The author gave a short report of what had happened at Amsterdam, and then summarized unspecified reports from other parts of the United Provinces, as well as Germany, France, and England. Interestingly, the report also mentions some speculations by "philosophers".[13]

In the end, most of this information is still based on letters, just as in the case of the earlier pamphlets. But unlike the pamphlets, newspapers keep to the facts as given by the letters, without an attempt at a theological or other interpretation. They served practical purposes. Merchants and regents wanted to know about events that might affect their affairs. Factual information was more relevant than edifying comments or spectacular wonders. "Secularisation" here lies more in the emergence of new public media, than in the emergence of new ideas.

This new emphasis on factual information appeared well in line with the demands of the educated public. In the second half of the seventeenth century, in many countries learned journals were established to inform the public about what was going on in the republic of letters. These journals became important sources of information in the learned world. After the earthquakes of 1692, Christiaan Huygens was curious what the *Philosophical Transactions* had on the subject and asked his brother Constantijn about it. Constantijn replied that he had not yet seen the Transactions, but he would inquire "whether those sirs had any special relation on the *seismòs* of Jamaica in order to let you know".[14]

The *Philosophical Transactions*, founded in 1666 in the slipstream of the Royal Society and later its official journal, had a whole series of earthquake reports. The journal published a report of the earthquake of Lima of 1687; several accounts of the Jamaican earthquake of 1692; and several reports of the earthquake of Sicily of 1693.[15] These accounts are strictly empirical. They are from eyewitnesses, try to be accurate and detailed rather than sensational, and refrain from any theological explanation. Not all reports were original. As was the case for newspapers, the journal also reprinted accounts published elsewhere. For example, it published an extract of the book by Burgos on the Sicilian earthquake of 1693.[16] Other learned periodicals published similar reports. The French *Journal des scavans* published a report of an earthquake felt at Paris in 1682.[17] The *Acta Eruditorum* published an extract from Salvadore Ruffo's report of the earthquake of Palermo of 1726.[18] The *Miscellanea curiosa*, the journal of the scientific society of the Holy Roman Empire, published observations and considerations on the Villach earthquake of 1690 and an earthquake in Slovenia in 1691.[19]

The Royal Society officially endorsed this emphasis on facts, making the factual and experimental approach a prerequisite for sound philosophy, thereby transforming the ideal of knowledge.[20] Whereas in the sixteenth century, works on history, nature, etc. focussed on marvels and rarities, in the seventeenth century people came to study such events as ordinary works of nature. Moreover, the Society had the ambition to function as a central point for the exchange and storage of factual information. They actively collected information. After the great earthquake of Sicily of 1693, they asked the Italian mathematician Giovanni Alphonso Borelli, who had earlier written on Mount Etna, to send a description. Borelli passed on the request to Domenico Bottone, professor of medicine at Messina, who had just published *Pyrologia*, a book on fire, including a large section on subterranean fires. (The book criticized Borelli's theory on Etna, by the way). Bottone complied with a long treatise which explained in detail the geographical and meteorological circumstances and the damage done in the various towns, but was completely silent about the theological interpretation. By the time the work was finished, interest in the subject in Italy seems to have subsided, but when years later a new earthquake occurred that struck Messina, it got printed with some additions on recent events.[21]

After the London earthquakes of February 8 and March 8 1750, the Royal Society discussed the topic at its meeting. An unusually large number of visitors attended. The Society also received a large number of letters about the earthquakes. They filled a complete issue of the *Philosophical Transactions*. The issue contained 57 contributions, most of them on the London earthquakes, a few on some other earthquakes felt in England the same year, one on an earthquake in France in 1749, one on earthquakes in Smyrna, and a few with more theoretical considerations.[22] As these last articles demonstrate, the speculative element was not completely lacking. Besides factual reports, the *Philosophical Transactions* did also publish more theoretical articles on earthquakes and their possible causes.

In other journals, this element may still be more prominent. The accounts in the *Miscellanea curiosa* are less factual than those in the *Philosophical Transactions*. They are not written by eyewitnesses who try to give an accurate description, but rather by scholars who base themselves on existing descriptions and try to turn those observations into a theory. One author discussed the movement of the earthquake in Tirol in 1670 with the movements in two of his patients and tried to establish a connection between the microcosmos and the macrocosmos.[23] The difference in character may be partly due to the fact that German scholars were less directly exposed to news from the overseas world than inhabitants of seafaring countries like England or the Netherlands.

Notes

1 Jorink (2010); Daston and Park (1998). Van de Wetering (1982) argues that eighteenth-century Puritan theology demanded a more rational interpretation of earthquakes.
2 *Erschröckliche Newe zeytung* (1542); *Zeittung* (1546); *Warhafftige newe zeitung* (1603); *Kurtze Erzehlung* (1630). Cf. also the fake report by Wettersteint (1673).
3 Cook (2007) 13–57, and *passim*.

4   Letter of 1616 October 26, cited in den Tex (1998) 45. The Gunung Api is presently called Banda Api.
5   T.K., *Journael* (1653) [A2].
6   T.K., *Journael* (1653) [A3].
7   Montanus (1669) 305–308, 353–354. Earthquakes are also mentioned on pp. 84, 103–106, 408. For an illustration of a volcano, see p. 416. den Tex (1998) 26–28 points out that Montanus' rendering of the facts is inaccurate.
8   Pettegree (2014) 182–207, 230–239, 245–248. Bogucka (1999). Cf. Schnurmann (2001) 254.
9   http://kranten.kb.nl/view/article/id/ddd%3A010759000%3Ampeg21%3Ap002%3Aa0005.
10 Wendeler (1691) C$^\text{v}$.
11 Drake (1996) 364. See also 321–322, on Hooke inserting in his lectures accounts from the Gazet on earthquakes in the West Indies in 1690.
12 *Terra tremens* (1670), no page numbers. On the Salzburg accident, see Hauer (2009).
13 *Europische Mercurius* III-3 (July-Sept 1692) 70–75. Cf. Heath (1692). Condorelli (2013) gives an overview of the news of the Sicilian earthquake of 1693 in European periodicals. See 141–142 for newspaper reports and 151, 157 for some examples of authors who used or incorporated the text of the gazettes in their own work.
14 Constantijn Huygens to Christiaan Huygens, 1692 Dec 30, in: Chr. Huygens (1888–1950) X, 380.
15 On Sicily: *Philosophical Transactions* no. 202 (July-August 1693) 827–829; nr. 203 (Sept. 1693) 893–894; nr. 208 (Febr. 1693/4) 2–10. On Peru and (mostly) Jamaica: *Philosophical Transactions* no. 209 (March-April 1694) 78–100. Cf. Willmoth (1987) 34.
16 *Philosophical Transactions* no. 202 (July-August 1693) 830–838.
17 *Journal des scavans* 10 (1682) 195–196 ('Relation du tremblement de terre arrivé à Paris et plusieurs autres endroits le 12 May 1682').
18 *Acta eruditorum* (1727) 39–45.
19 Schelhammer (1691); Thalnitscher (1691).
20 The history of the concept of fact in science is complicated, but I cannot elaborate here. See Shapiro (2000).
21 Condorelli (2013) 159–161; Bottone (1718).
22 *Philosophical Transactions*, vol. 46, issue 497 (1750 Dec 30). J.G. Taylor (1975) 15–16.
23 Dobrzensky (1671).

13

NEW OBSERVATIONS AND THEORIES

The non-confessional investigation of nature

For centuries, natural philosophy had been a part of an overall attempt to understand all aspects of reality. In the climate of the sixteenth century, this meant that it was closely aligned with established religious orthodoxies. By the middle of the seventeenth century, however, the situation changed. Many scholars started to study the phenomena of nature deliberately ignoring any confessional framework. It was now felt that all phenomena could be reduced to some immutable and universal "laws of nature". Questions about the significance or prodigious nature of specific phenomena were banned from natural philosophy.

This shift found its justification in new philosophical principles, but that does not mean that those principles actually led the way. It appears more likely that the "new philosophy" sprung from, and served to legitimate, an approach to nature that had already gained traction earlier in the seventeenth century, when the enhanced means of communication and the greater access to information had given rise to a new, fact-based approach to the world. This fact-based understanding of the world was a necessary precondition for the rise of new mechanical ideas of nature, but empirical research of the facts of nature was not dependent upon a specific philosophy. One of the most important "empirical" researchers of the seventeenth century was Johan Baptist van Helmont whose careful observations, as we saw, were made in spite of his being a follower of a spiritualist, Hermetic philosophy.

The study of earthquakes and related phenomena was changed not so much by the "new philosophy", but by what I call the "new empiricism". Although the approach was for the moment not particularly successful in helping to understand earthquakes, it did assist in introducing a new way of studying nature. So, it would seem that the so-called seventeenth-century "scientific revolution" was not just a consequence of developments in mechanics, astronomy, optics, anatomy, and physiology, but that the study of the earth played some part as well.

The new empiricism of the seventeenth century

The term "new" empiricism definitely is not meant to imply that earlier people were not interested in concrete data. That of course is not true. People collected information all the time. In the case of earthquakes, governments collected information on damages for the purpose of tax exemptions or military defence. Authors compiled lists of earthquakes and tried to relate them to other events. Some scholars collected information on an earthquake in a systematic way and made observations not just on the destructions, but also on atmospheric conditions, meteors, and so on. The sciences as they developed in the sixteenth century were in certain respects quite empirical. What changed in the seventeenth century is not so much that people started to collect data and make observations, but rather that they created a new framework in which such observations obtained their meaning.

The problem was to decide what facts were relevant and what did they indicate. Before the middle of the seventeenth century, people looked for facts that helped them understand the will of God and the course of world history. Earthquake catalogues served to demonstrate that earthquakes were forebodes of evil, not to find a geological pattern. A new generation of philosophers however came to seek the origin of all phenomena in natural causes only, unrelated to human affairs. Earthquakes now were discussed in connection to the constitution of the earth and the origin of mountains, not the sinful state of man or the impending Last Judgement. In earthquake histories, the physical world became more the focus of attention than human reactions or the whims of fate. Attention gradually shifted from demolished buildings and personal stories to discussions of the weather, the soil, and other natural particularities, studied in a physical, not a theological context.

We can see this, for instance, in what people write on the behaviour of animals. People had taken notice of animals during an earthquake for a long time, but by the late seventeenth century, the reason to do so had changed. Renward Cysat in his notes on the earthquake of 1601 wrote in a rather detailed way on the behaviour of dogs in the streets, the cattle on the mountains, the rats in the houses, even the insects. However, he did so not out of an interest in zoology, but to underline the religious meaning of the event: the animals behaved "as if they expressly noticed the anger of God".[1] On the other hand, when in 1737, the physician Textor recorded the behaviour of chickens during the earthquakes at Karlsruhe, he did so out of interest in the terrestrial exhalations that supposedly caused the abnormal behaviour.[2] Cysat may have been the better observer of the two, but he did not use his observations as clues to understand nature, whereas Textor self-identified as a natural investigator.

More complicated is the case of the thin clouds or specific meteors that were seen before an earthquake. Already Aristotle had regarded those as indications of the natural causes of earthquakes and many philosophers in the Middle Ages and sixteenth century followed suit. But at the same time, many of these phenomena could be seen as ominous. Fiery meteors were traditionally seen as portents. The same is true for the smell of sulphur that some people reported to have sensed. There were natural explanations available, but for others, the smell of sulphur indicated the activity of the devil.

By the second half of the seventeenth century, however, the attention for the weather and other environmental conditions took on an unambiguously naturalistic character. Not that the signs that Aristotle had given for impending earthquakes, like thin clouds, a calm atmosphere, and the like, were dismissed. If anything, they got more attention in the new empirical climate. An account of the weather and the atmospheric condition at the moment of an earthquake remained a standard element in the descriptions. Aristotelian elements like the quiet atmosphere, the long thin cloud, and the smell of sulphur are mentioned well into the eighteenth century. Stephen Hales still referred to them in 1750 as phenomena which gave a clue to the cause of earthquakes.[3] Of course, it is only with hindsight that people remembered to have seen such clouds and one may wonder how reliable their memory was, especially given the fact that such signs were more or less expected.

By the middle of the seventeenth century, philosophers started a conscious programme to explain many phenomena that were traditionally seen as portents, as ordinary works of nature. This was a matter of fierce controversy. Best known are the debates on the significance of comets that were waged on the occasion of the comets of 1665–1666 and 1680, but these were not the only issues at stake. Monstrous births too were increasingly regarded as common events that demanded a medical, not a theological explanation.[4]

In course of time, this debunking of prodigies became heavily associated with mechanical, in particular Cartesian philosophy, but it had a much wider appeal. In 1647, the Flemish priest Govaert Wendelen published at Brussels a treatise that investigated a purple rain that had fallen in the city the year before. After a short description of the event, the volume was mostly filled with a correspondence between various scholars on the natural causes of the phenomenon. Involved were Pierre Gassendi, Vopiscus Fortunatus Plemp, and several others, none of them Cartesians and apart from Gassendi none of them really mechanical philosophers. The main tenor was that the rain was a purely natural phenomenon, not some kind of prodigy. In 1689, the German physician Philipp Jakob Hartmann published an article in *Miscellanea Curiosa* wherein he discussed a rain of grain. He argued that all such phenomena should be explained from natural causes, but did so along medieval-scholastic lines.[5] It is definitely true that Descartes and other mechanical philosophers gave a philosophical legitimation to the study of the world as a self-contained whole. He strengthened the movement and gave it a programme. Its origin, however, lies in the increasingly physical, rather than moral view of the world, which probably owes as much to the creation of new intellectual and institutional networks and conventions of communication, as to new philosophical principles.

In the eighteenth century, fiery meteors were no longer regarded as signs which indicated a divine message, but purely as indications of the physical cause of the quake. Remarkable is an English relation of the earthquake at Lima of 1746, wherein it is stated that the inhabitants, "being very superstitious", interpreted a fiery meteor seen in the night of 14 May as presaging an impending terrible earthquake. Even the fact that the earthquake indeed did occur only brought the author to the admission that "it was a right Conclusion drawn from wrong Principles".[6]

The tendency can be discerned in works of a more religious inspiration as well. Salvatore Ruffo's description of the earthquake of Palermo in 1726 is in a sense typical for the confessional tradition, full of stories of miraculous savings, conversions, penance, and processions. But Ruffo also included some more physical considerations. He asserted that the parts of the city where the destruction was heaviest were those where in ancient times the two harbours had been. "For although the buildings there have solid foundations, because they stand on a chalky and sandy soil, a remainder of the ancient harbour, they are still more liable to be shaken and destroyed by an earthquake."[7] The mathematician Georg Friedrich Richter, who published both a Latin and a German translation, agreed and in his edition even included a map of the city with the ancient situation.

Ruffo also paid attention to the meteorological situation that preceded the earthquake. He mentioned in particular a dark cloud full of lightning that had been observed on a French ship the night before, and that ended with a terrible lightning stroke that upset people in the city.[8] The description of this meteor may well have been one of the main reasons for Richter to translate the book, as he felt that it supported his own theory on the origin of lightning. Richter theorized that both lightning and earthquakes were caused by the kindling of an inflammable *halitus* from the earth, consisting of particles of sulphur, nitre, copper water, alum, and the like. In the case of lightning this happened above the earth, in earthquakes below. He was engaged in a polemic with the Italian scholar Scipio Maffeo on the topic and was happy to get some extra arguments. He felt it important to have Ruffo's work translated, for

> it is sufficiently known how useful in physics are the histories of natural effects that have been transmitted faithfully and accurately. Not only new observations are helpful, but old ones too have to be confirmed often and repeatedly, so that safely supported by these, we are able to notice also circumstances that are often mistakenly ignored.[9]

Both the shift in outlook and its limitations can be illustrated by the case of Johann Christoph Sturm, professor and a prominent philosopher at Altorf near Nuremberg. He kept aloof from the dominant philosophical schools and preferred an experimental approach. Indeed, he was among the first university professors to illustrate his lectures with actual experiments. In his *Philosophia eclectica*, he published a number of previously held disputations. Many of these are on general subjects, but several discuss specific topics. The eleventh disputation, originally defended on 24 September 1670 by a certain Jacob Hieronymus Lochner, is on earthquakes. It was held on the occasion of a large earthquake in July of the same year, which was felt in a large part of southern Germany and northern Italy.

Sturm starts with a declaration of principle. In order to find the cause of earthquakes, he will not occupy himself with etymologies or the rehearsing of other philosophers' opinions. The true method of the investigation of nature, applied to such great success in recent times, consists in collecting information on the specific effects and circumstances, based on indubitable experiences. The data have to be carefully compared, in order to identify the cause which is responsible for the whole.

Sturm than continues by listing thirteen *Phaenomena* that accompany earthquakes. After the earlier declaration of principle, these may sound somewhat disappointing. Sturm inevitably has to rely mostly on observations and accounts by others. Even though he seriously tries to sift his material critically, many of the phenomena he lists are simply those familiar from ancient literature: the various types of earthquakes; the fact that earthquakes occur, especially in mountainous areas; the signs by which they are preceded (muddy water in pits, nebulosity around the sun, etc.); and the effects by which they are followed – plague, hunger, and rebellion. However, Sturm is not following the ancient authorities slavishly. Other *phaenomena* he mentions run counter to received ideas or are based on more recent observations: contrary to the scholastic tradition, he recognizes that earthquakes occur in all seasons and under all astronomical constellations; and he states that after a large earthquake during which fire has erupted, the next one will normally occur only after a long interval, mostly coinciding with extinct volcanoes becoming active again.

These thirteen phaenomena are followed by eight "observations". These are general observations on the constitution of the earth: there are many cavities, subterranean canals, etc.; there is much sulphur, saltpetre, bitumen, and metal within the earth, especially at places where earthquakes normally happen. Sturm concludes that earthquakes are caused by the explosions of subterranean exhalations in the way of siege mines. A long list of conclusions argues how the various phenomena can be explained by this hypothesis. A final chapter rehearses the opinions of some earlier, mostly seventeenth-century, philosophers.

To us, such a survey of bookish learning may not appear very "empirical". Sturm could feel justified to blaze a new trail not so much because of the way he collected his data, but because of the way he regarded them. The causes of earthquakes were physical and were to be discovered from the phenomena, that is, from facts in nature. There is not the slightest attempt to interpret earthquakes in a religious framework. Sturm keeps his physics rigorously distinct from theology. Nor is there any attempt to see the various occurrences in another way than as physical phenomena.

Volcanoes, earthquakes, and subterranean conducts

The geographical information that was pouring in with the voyages of discovery transformed the view of the earth. Although the ancient Greeks and Romans knew about what we now call volcanoes, notably in the case of Etna, they did not classify these as a specific phenomenon. But when European seafarers swarmed out over the globe, they met with similar eruptions in many other places. As a name for the phenomenon was still lacking, Spanish or Portuguese seafarers introduced the term *vulcan* or *volcan* for them, after the island of Vulcano in the Mediterranean, one of the Aeolian islands. Scholars compiling the information from travel reports soon realized that on a global scale, such phenomena were not particularly rare. Johan Baptist von Helmont in 1640 made a clear distinction between volcanoes and earthquakes – the former were natural, the latter supernatural. The first list of the world's volcanoes dates from 1650 and was published in Bernard Varenius' *Geographia generalis*. Varenius introduced the term "volcano" to the learned world. He even distinguished between active and extinguished volcanoes.[10]

Other authors were to follow. In order to demonstrate that the earth was full of fires, Athanasius Kircher gave a list of the world's volcanoes in his *Mundus subterraneus* of 1665.[11] The German professor Thomas Ittig in 1671 published an ambitious 340-page work on "burning mountains", that is to say, volcanoes. A first section gave a catalogue of all the volcanoes in the world. Italy and the rest of Europe still got most attention, but there were chapters on Africa, Asia, and America as well. The second section dealt with the opinions of earlier authors. In a third section, Ittig attempted to come up with the true explanation, wherein subterranean fires took a prominent place. A fourth section discussed various phenomena that accompanied volcanic eruptions.[12] Whereas the earlier catalogues of earthquakes or other prodigies were mostly attempts to understand world history, that is the history of Christianity, such catalogues of volcanoes were purely geographical. Volcanoes thereby became regarded as normal natural phenomena. This could not fail further to promote ideas on subterranean fire.[13]

Academic works on volcanoes are of course the culmination of a long process. Speculation on volcanic phenomena already started in the first half of the seventeenth century. A watershed in the debate on subterranean fires was the devastating eruption of Mount Vesuvius in 1631. Vesuvius had been dormant for many centuries. When it finally awoke, its unheard-of vehemence and power greatly impressed contemporaries. People were familiar with the periodic eruptions of Etna or Stromboli, but those dwarfed in comparison to the torrents of fire that Vesuvius now produced. An endless stream of pamphlets, news-sheets and other publications testifies to the awe of the contemporaries.[14] Natural philosophers were faced with the challenge how to explain this unheard-of phenomenon. Quite naturally, they fell back on the idea of subterranean fires. But whereas earlier thinkers had mostly seen these fires as strictly local, even though authors like Agricola and Cardano had entertained the idea of subterranean conducts in which burning material could move from place to place, seventeenth-century philosophers were considering them on a more global scale.

One of the first recorded attempts to explain the eruption of the Vesuvius happened in the circle of the famous French erudite Claude Fabri de Peiresc. In or shortly after 1632, Peiresc was visited by a Capuchin friar, Aegidius de

Losches, who had just returned from Egypt. The reason for his visit was a common interest in oriental manuscripts. However, De Losches also reported on the eruption of Mount Semus, a volcano in Ethiopia, which had happened at the same time as the eruption of Mount Vesuvius – the kind of information that would have been very hard to come by in earlier times. Peiresc and Losches felt that the simultaneity was no mere coincidence. They started to talk "about the subterranean conducts, by which not only waters, but also fires could be communicated". In this way, they felt, Vesuvius could be connected to Etna, Etna to Syria, Syria to Arabia, and Arabia to Erythrea, where Semus was situated. That the subterranean fires could force themselves a way was indicated by the creation of new mountains, as the one at Pozzuoli in 1538.

> That the covering crust is sometimes disrupted, so that water pours in, is indicated by the fact that at the time of the eruption of Vesuvius, the coastline at Naples withdrew for a while, whereas the mountain revomited the water that had been absorbed through the openings, though mixed with combustible and burning material.

The fire that had earlier erupted from Etna could be interpreted in a similar way.[15]

The idea of subterranean conducts wherein some form of fire circulated was eagerly taken up.[16] Peiresc himself never published these ideas, but Pierre Gassendi, a long-time protégé and prominent member of the circle of French *libertins*, included them in his biography of Peiresc. Though buried in this big volume, the observation drew attention and was eagerly referred to in other works. Other people came up with similar ideas at about the same time. Agatio Di Somma in his description of the earthquake of Calabria in 1638 did not doubt the existence of hidden sulphurous veins. At the surface of the earth, the trajectory of these veins was indicated by sulphur pits, bitumen, cracks in the earth, and at the time of the disaster by the odour of sulphur. Based on these indications, Di Somma posited a connection between Calabria and the island of Vulcano. In his view, the enhanced volcanic activity on Vulcano in 1638 indicated that the disaster that struck Calabria in 1638 probably had its origin there.[17]

The theory of the subterranean fire that circulated through the earth was used to explain other earthquakes as well. Bottone in his history of the earthquakes of Sicily of 1693 noticed that the main earthquake seemed to have its origin in the south. Malta was struck earlier than Sicily, the south of Sicily earlier than the north. He explained this by assuming that the movement followed long subterranean conducts, wherein some *halitus*, generated from minerals, accumulated and was ignited. He noted that Kepler and Cabeo had felt that the veins in the earth run mostly north-south and east-west, and that the present earthquake seemed to confirm that. Bottone suspected that a sulphurous spring at Castra Regia on Sicily was connected to a similar spring on the island of Lipari via a canal that went under the sea. He noted that the 1693 earthquake abated and returned to a quiet stage after an eruption of Mount Etna. This apparently had given an outlet to the fire that had earlier been confined within the earth.[18]

The Sicilian earthquake gave rise to many similar theories. A disputation at Jena speculated that this earthquake found its origin in the fact that the vents of Etna had been blocked, so that the subterranean fires could no longer escape.[19] The theory of subterranean conducts was also advocated by the Italian physician Giorgio Baglivi and the English botanist John Ray. Baglivi even saw a connection between the earthquake of Sicily and one simultaneously taking place in Peru.[20] Still in 1746, the German physician and philosopher Johann Gottlob Krüger maintained that earthquakes had their main seat in Naples and Sicily, that is, near Etna and Vesuvius. He was vague about the exact cause of the subterranean fires, but these heated the air in the caves so that it expanded and either shook the earth or made the vault collapse, so that the fire would erupt.[21]

Subterranean fires were not just discussed as explanation for specific phenomena, like hot springs or exhalations. Philosophers picked up the idea of a subterranean fire and incorporated it into their more general theories. From the 1630s onward, the subject increasingly was discussed as a topic in its own right. So, in 1641, Giovanni Nardi published in Florence his book *De igne subterranea physice prolusio* [Physical prelude on the subterranean fire]. The professor Heinrich Rixner had a disputation on subterranean fire defended at Helmsted in 1656. It was, for a disputation, a rather ambitious work of thirty pages. Another disputation on the fire in the sky and the earth was defended at Duisburg in 1698. Such discussions of subterranean fire would include sections on phenomena like earthquakes, but these were no longer the major part. In this way, the observations of volcanic phenomena contributed to new ideas on the constitution of the earth.[22]

There was criticism as well. Not everybody believed in the central role of fire. Giovanni Alphonso Borelli, who wrote a pioneering work on the eruptions of Etna, was quite sceptical about it. But although his work was founded on sound empirical research, it could not impress the proponents of subterranean fire. Bottone criticized Borelli's arguments in his *Pyrologia*.[23] Another critic, Giovanni Bottari, felt that the story of Mount Semo was by hearsay and hard to check. Anyway, the simultaneousness of those events could have been accidental. He also referred to Borelli's researches on Etna. Bottari preferred to think of the earth as a solid body. Even if there were subterranean connections between the various volcanoes, these could hardly be the cause of earthquakes.[24]

In eighteenth-century earth sciences, volcanoes and subterranean fire typically had no prominent place.[25] However, for the thinking on earthquakes these developments were of great importance. The idea that earthquakes were due to some subterranean fire that would move from one place to the other through subterranean canals seems to have quickly taken hold of popular imagination. A German popular chronicle that was published on the occasion of the 1692 earthquake, talks about earthquakes as a phenomenon that has its origin in a specific locality, and then moves on. The author tracked the movement of the 1692 earthquake and also maintained that the earthquake of 1580 had had its origin in the Low Countries, from whence it went partly to Paris and partly to England, where it reached York.[26]

The most striking notion of the idea is to be found in an English pamphlet of 1673, which purports to give a report on an earthquake in the Ottoman Empire, but is completely made up. The author apparently wanted his story to be spectacular but at the same time credible, and therefore took care to be in agreement with accepted notions. The supposed quake lasted many days, meanwhile moving from place to place; it caused successive fissures in the earth and eruptions of fire in many different places. The author concluded that it was caused "by an impetuous spirit or blast of wind included in the bowels and caverns of the earth, which ran so many hundreds of miles through the blind caverns of the earth, before it could get vent".[27]

The motion of the earth

Earthquakes, people realized, gave indications of the inner constitution of the earth. But some people also felt that they offered a clue to the behaviour of the earth itself. Especially in Italy, where the Copernican theory of the motion of the earth was still banned, some Catholic scholars were eager to find a physical demonstration of the motion of the earth that could annul the religious objections, and to that purpose looked at earthquakes.

One author who hoped that earthquakes could prove the Copernican theory was Francesco Travagini. His idea was that earthquakes eventually were caused by the rotation of the earth. He had gotten this idea when during an earthquake in Emilia Romagna on 1661, April 22, he observed that the tremors were all in the east-west direction. When in 1667 he was in Venice during the great earthquake of Ragusa, he made observations to confirm his hypothesis. He noticed that the tremors were all from east to west; that in the canals, waves were formed running east-west; that the clock towers and other high buildings moved in the east-west direction, so that only the buildings standing east and west of them were damaged; and that free-swinging objects, like church lamps, did so from west to east. He had noticed these same effects at the earlier earthquake in Emilia "as I read in the diary of my observations".[28] If the east-west motion of earthquakes were a general rule, it seemed obvious that this could only be caused by the rotation of the earth.

A similar theory was defended by the Polish Jesuit Adam Adamowich Kochanski. Kochanski referred to an earthquake on May 17, 1667, which he had experienced when he was a teacher at the Jesuit College at Florence. He too had been struck by the fact that the tremors seemed to move from east to west, and instantaneously realized that if this was universally true for all earthquakes, this phenomenon might be fit for proving the daily rotation of the earth. He immediately asked Jesuit colleagues who had experienced earlier earthquakes for corroborating information. Later, he collected information on other earthquakes.[29]

In the end, the theory that there was a connection between earthquakes and the rotation of the earth did not get much support, although there were still sporadic references to it in the eighteenth century. Cirillo in his report of the earthquake of Apulia of 1731 claimed that earthquake oscillations tended to move along the earth's parallels and that this was a strong argument for the daily rotation of the earth.[30] An academic disputation defended at Utrecht in 1747 referred to Kochanski's work and even stated that one of the main uses of earthquakes was that they provided astronomers with an invincible argument for the daily rotation of the earth.[31] To most observers, however, it must have become clear pretty soon that the east-west movement was no general rule.

What the work by Travagini and Kochanski did accomplish was that it made people aware that the direction of the tremors might be a topic worth investigating. Indeed, it became very much a standard point in earthquake observations. The Dutch physicist Nicolaas Hartsoeker in his *Principes de physique* referred to the earthquake of 1692 which he had experienced at Amsterdam. Among other things, "a pendulum of about three feet long, which was suspended from a wooden attic [*au plancher*], continued its movement for at least half a quarter of an hour, swinging pretty closely from north to south".[32] Christiaan Huygens noted that during this same earthquake at Amsterdam, the waves had moved from north–north–west to south–south–east.[33] Regrettably, he does not tell who made this remarkably precise observation (he himself was near The Hague), but clearly, some people were taking notice.

Mapping strength and direction

This idea that earthquakes were caused by some fiery substance moving in subterranean conducts, became one of the main motives for making accurate observations of earthquakes. Scholars relating earthquakes to the eruptions of distant volcanoes and other phenomena realized that in order to demonstrate the connection, they had to compare the times that the various events happened. The idea itself was not new. Already in 1580, Thomas Twynne guessed that the earthquake in England had moved from east to west, since the destruction in the east was greater. He corroborated this idea by checking at what local time it had been felt at various places. Since clocks were still pretty inaccurate, his estimates appear very rough, if not inspired by wishful thinking.[34]

In the seventeenth century, ever more accurate registrations became possible. The description of the physical circumstances of earthquakes became much more detailed, including the time, the places where it had been felt, the time it lasted, and the direction of the tremors. To establish the condition of the atmosphere, people no longer just looked at the sky, they studied the (newly invented) barometer. Christiaan Huygens, the inventor of the pendulum clock, felt the earthquake of 18 September 1692 in his country seat of Hofwijck near The Hague at 2:30 p.m. He noted that the earthquake lasted ten to twelve seconds and made precise records of the changes in the barometer. He noted that it had been felt in many other places, most severely in Liège. "There the time was 2:15 p.m., so earlier than here."[35] The physician Textor had his report of a series of earthquakes at Karlsruhe in 1737 accompanied by detailed notes on the time, the weather, and the readings of the barometer (of which he gave the division) at every quake. He also made observations of accompanying sounds, meteors, and (as already stated) the behaviour of animals. He even excused himself that his watch did not indicate seconds and that because of other obligations he had not been able to make continuous registrations.[36]

Identifying the source and direction of earthquakes became an important goal of these observations. As we saw above, Domenico Bottone felt that one could track the movement of the earthquake of Sicily in 1693 by the times that the various lands shook. John Ray collected information on the 1692 earthquake as it had been felt in England. He had not noticed the quake himself and could get no exact information from the Essex village where he lived, but at London, it had been felt at four minutes past two p.m., lasting about two minutes. "Had we punctual and exact Notice of the very Minute that it happened in far distant Places, we might thence gather something concerning the Motion and Progress of it." Ray also realized that if we could identify the place where the earthquake had its greatest force, we might deduce where its first "Accension" was and which way it spread itself.[37]

Such comparisons were possible only because of better instruments and improved means of communication. It was mainly for the purpose of tracking the origin and movement of earthquakes that the first instruments made for measuring their power and direction were introduced. Travagini and Hartsoeker had already watched the movement of pendulums during an earthquake, but these do not appear to have been suspended for that very purpose. A deliberate use of pendulums for measuring earthquakes happened only in the eighteenth century. The first case we know of is a report by Nicola Cirillo, physician to Pope Innocentius XII, on an earthquake in Apulia in 1731, sent to the Royal Society. Cirillo noted that the earthquake had been felt strongest in the village of Foggia, and proportionally weaker at greater distances. "The same has been attentively observed in the swinging of pendulums placed at various distances from Foggia, by two observers who had agreed to devote themselves to this." The pendulums were a palm long and had a scale that indicated the amplitude of the swings. One was at Asculo, the other at Juvenato. Cirillo was happy to note that the difference in amplitude appeared inversely proportional to the square of the distance to the presumed centre, Foggia.[38]

Another pendulum for earthquake observations was introduced in 1751 by Andrea Bina, a philosophy teacher of the Benedictine order in northern Italy. In a book which was mainly devoted to physical speculations about the causes of earthquakes, he also included a description of an instrument to record earthquakes: a heavy pendulum with a pin that left a trace in a box of wet sand underneath.[39] His pendulum therefore was not just a measuring, but also a primitive recording device, although it would seem a bit overdone to call Bina the inventor of the seismograph. In the second half of the eighteenth century, several other ideas were floated and instruments underwent further improvements.[40]

A different kind of instrument was proposed by the French priest Jean de Hautefeuille, an eager amateur scientist and inventor of instruments. In 1703, he published a booklet with various newly invented instruments. Among these was "a means to make observations on earthquakes, and be able to predict them", as the title page claims. The instrument consisted in a bowl filled with mercury, with a broad rim in which were eight cavities or vases. In the case of an earth-quake, the mercury would spill over into one or more of the cavities; the one that was filled indicated the "inclination of the earth" and the direction where it had started to rise. The force of the earthquake could be determined by meas-uring the quantity of spilled mercury. Hautefeuille suggested that a series of his instruments would give a better idea of the strength and origins of an earthquake:

> If one takes this experiment in several places, one will know the origin of the quake and the place where it has started: because since the earth will have been lifted there on all sides, the mercury will fall in the cavities in different directions.[41]

The same principle had been used by the Chinese already more than a millen-nium earlier, but it is not clear whether Hautefeuille was aware of that. There are no indications that his instrument was ever built. Still, this type of instrument, now called a seismoscope, was considered by several other eighteenth-century scientists. A similar seismoscope, but now filled with water, was built by Chan-dler in Lisbon in 1742. A very elaborate instrument, connected to a timekeeper, was put up in 1784 by the abbé Atanasio Cavalli at the *Specola Caetani* in Rome and actually used for observations.[42]

For the moment, these instruments had little impact. They remained isolated episodes or loose ideas by relatively obscure men. Leading scientists were not involved and none of the major learned societies promoted an effort of systematic data collection by means of such instruments. Such efforts had to wait until the nineteenth century. However, the fact that such ideas were discussed at all testi-fies to the growing awareness of the importance of more and better empirical ob-servations. It would seem that the ideal of systematic data collection had existed for a long time among the educated public before the scientists finally took it up.

A plethora of theories

With a new programme for natural investigation in place, and not allowing themselves to be constrained by confessional or didactic demands, the "new phi-losophers" of the seventeenth century had a fresh look at many traditional prob-lems. One of these was the cause of earthquakes. If they did not come up with any really convincing answer, it was not for want of trying. The phenomenon was simply too intriguing to be left alone. As the Aristotelian principles had been discarded, scientists had to find alternative explanations. Cartesians and other mechanical philosophers, advocates of British experimental philosophy, eclectics, and chemists, they all came up with new suggestions.

The main innovation in natural philosophy of the seventeenth century was arguably mechanical philosophy. In particular, the philosophy of René Descartes, which offered a complete alternative to Aristotelianism, was immensely influential. Besides new theories of motion, light, the universe, living beings, magnetism, and the soul, Descartes also offered a new theory of the earth. It would be the point of depart of many later discussions on the constitution and history of the earth. A major principle of Descartes was that there was no essential difference between the terrestrial and celestial worlds. The earth had originally been a star. Stars were bodies of fiery material that were the centre of a big vortex of celestial matter. They had constantly to purge themselves from coarser particles that concreted on their surface, as was visible in sunspots or variable stars. In some cases, the coarse matter in the end formed a solid crust. The vortex around the star then collapsed. When the now dark celestial body was caught in the vortex of another star and started moving around it, it had become a planet, like our earth.[43]

Consequently, the centre of the earth had still a fiery, starry, nature. Quite likely, Descartes came to this theory by freely reusing existing ideas on subterranean fire and earth history, but unlike the Jesuits to be discussed later, he avoided any hint at the biblical world view. He gave a detailed (though hypothetical) description of the constitution of the earth. Based on his theory of the earth's origin, he described how in course of time around the fiery core various layers formed of air, water, and stony and metallic matter. The sulphurous and nitrous vapours that according to common understanding caused earthquakes got their place in this model as well, so the common theory of earthquakes was easily incorporated into his system. Descartes' theories offered a daring synthesis, but they were largely put together from existing elements, including existing ideas on subterranean fire.

As a consequence, his new philosophy contributed very little to the understanding of earthquakes. Most mechanical philosophers simply took the old theories and now described them as brought about by mechanical principles. They still attributed earthquakes to subterranean sulphurous and nitrous vapours, which were kindled and exploded. This theory, as we have seen, was already quite popular in the Renaissance. The one difference is that the production and explosion of the vapours was now interpreted as a mechanical process, whereas earlier authors had described the sulphurous and nitrous vapours in terms of Aristotle's dry exhalations.[44]

Arguably the most influential earthquake theory based on mechanical principles was offered by the French chemist Nicolas Lémery, a member of the Paris Academy of Sciences. It was mainly based on his experience in chemistry. Lémery gave a new turn to the by now centuries-old idea of subterranean fire by abandoning the Renaissance idea that the inflammable exhalations had their origin in sulphurous, nitrous, and bituminous substances. Instead, he proposed a mixture of sulphur and particles of iron. The idea was first put forward in his influential textbook *Cours de chimie* (1675). Continuing this line of thought, in 1700, he performed an experiment to investigate and illustrate this idea. It was published under the title "Physical and chemical explanation of subterranean fires, earthquakes, tempests, lightnings and thunder" in the *Mémoires* of the Paris Academy of Sciences for the year 1700.[45]

The basic observation was that when one takes equal parts of powder of sulphur and iron filings, mixes them with water to a paste, and lets this paste stand for two or three hours, the mixture will get hot. The paste will burst, hot vapours will escape; in the case of a large quantity of matter (thirty to forty pounds), these vapours will even catch fire. This, according to Lémery, gave a most likely explanation of the fermentations, violent motions, and fires that one could observe in volcanoes like Etna or Vesuvius. Lémery conducted several experiments with this mixture. Most famously, he buried a pot with a quantity of fifty pound of this paste a foot deep within the earth. Eight or nine hours later, "the earth swelled up, became hot, and got cracks; next, sulphurous hot vapours escaped, and then some flames that enlarged the openings and spread a yellow and black powder around the place".[46]

Lémery concluded that earthquakes were caused by a vapour that was produced by this fermentation of iron and sulphur. Just like earlier generations of philosophers, he felt that these reactions took place in caves or hollow places, arguing that for such reactions air was needed. The vapour then was converted into a sulphurous wind that forced itself a passage, "lifting and shaking the lands under which it passes".[47] If this wind was constantly strengthened without being able to escape, the result was an earthquake that would last until the wind had lost its motion. If it found an opening, it rushed out violently, which would result in a tempest. Lémery explained a number of other phenomena from his theory: volcanoes, hot springs, will-o'-the-wisps, thunderstorms, and thunderstones.

The "volcano of Lémery" for a long time remained a standard element in the explanation of earthquakes. An academic disputation from Gdansk from 1728 discussed Lémery's theory quoting the description in the German physics of Christian Wolff. The author, the physician Johann Adam Kulm, felt that the experiment could account for all the phenomena of earthquakes, whereas the theory of a permanent subterranean fire made no sense.[48] The mathematician Johann Bernoulli praised Lémery's theory in a letter of 1737.[49] The famous *Encyclopédie* by Diderot and d'Alembert also mentioned the experiment, but added a criticism by the chemist Rouelle, that the theory implies that iron exists in the earth in pure form (rather than ore), which seems improbable.[50] Lémery's experiment was still described in 1790 in an encyclopaedia of the sciences, where it was explained that the same phenomenon on a larger scale undoubtedly would cause earthquakes and volcanoes.[51]

To us, his explanation of earthquakes in terms of "sulphurous winds" sounds rather traditional. However, Lémery was a follower of Descartes. He felt that all chemical reactions followed from mechanical interactions between particles. The reaction of sulphur and iron is described in chemical turns as a "fermentation", but this fermentation is brought about "by the penetration and violent rubbing that the sharp points of the [particles of] sulphur make against the particles of iron".[52] In the end however this underlying idea had little practical consequences. It is just a new dress for an old theory. A similar jumbling of traditional theories with the new corpuscular view of nature can be found in other mechanical philosophers as well. (The philosopher Claude Gadroys even gave a mechanical explanation of astrological predictions.)

Lémery was not the only scientist who drew upon his chemical expertise. Just as in the sixteenth century, people had found an analogy to earthquakes in the effects of gunpowder, in the seventeenth century new chemical experiments and insights served the purpose. In Britain, Martin Lister got some renown for his theory of pyrites. His description of pyrites was based on chemical analysis, but from there he moved to further speculations on the role of pyrites in the economy of nature. Pyrites, he claimed, generated a hot, sulphurous, vehement, and inflammable *halitus*. The effects of this *halitus* are lightning when kindled in the air, earthquakes when they are kindled in subterranean caves, and hot springs when they are carried by underground streams.[53]

A similar theory was taken up by Stephen Hales in a paper read before the Royal Society after the London earthquake of 1750. Hales too referred to chemical experiments. He had found that when pure air was mixed with the sulphurous vapour made from pyrites and nitric acid, this resulted in a violent agitation and a reddish fume, of the same colour that (he claimed) was seen in the air before earthquakes. From this, he argued that exhalations of sulphuric vapours from the earth, when mixed with the atmosphere, could cause lightning, earthquakes, and cyclones.[54]

The English author John Woodward speculated that earthquakes were caused by water within the earth that, heated by the subterranean fires, evaporated, expanded, and when it got blocked caused a heavy commotion.[55] The professor Georg Christian Maternus de Cilano, in a disputation defended at the gymnasium of Altona (near Hamburg) in 1741, rejected the idea that earthquakes were caused by inflammable materials like sulphur, saltpetre, or bitumen. He felt that they were caused by the expansion of compressed air. In a sense, he went back to the ideas of Aristotle, but now understanding wind as a form of air and referring to the modern theories of pressure of the air.[56]

There are also completely different theories.[57] The French scientist Guillaume Amontons presented a paper in the *Histoire* of the Paris Academy of Sciences in 1703 wherein he defended the opinion that earthquakes had their origin in the effects of heat on the air deep inside the earth, which had been compressed to the density of gold.[58] The English astronomer royal John Flamsteed, in a letter from 1693, speculated that earthquakes were actually caused by explosions not in the earth, but in the air, that is, the atmosphere.[59] Flamsteed did not publish his letter, but the "airquake" theory was a few decades later publicly and elaborately defended by the Italian physician Giovanni Bottari.[60] The French military officer and traveller Amedée François Frézier argued that earthquakes were caused by water that made the soil underneath unstable, a theory that elicited some controversy.[61]

Interestingly, in the climate of theoretical uncertainty, even astrology was sometimes called upon. Most systematically this was done by Flaminio Mezzavacca, a doctor of law from Bologna, in his treatise on the earthquake of 1672. On the whole his treatise is fairly traditional, but when he comes to speak of the signs that announce earthquakes, he includes a discussion of the influence of the heavens. He gives the horoscopes (partly taken from other sources) of eight major earthquakes between 1570 and 1672 and from these tries to draw conclusions what configurations are particularly dangerous.[62] As late as 1748, a report of the great earthquake of Lima opened with a detailed description of the celestial configuration at the moment of the quake, describing it as fatal for the particular region of Lima and clearly seeing it as the direct cause of the event.[63]

Later in the eighteenth century, the experiments by Benjamin Franklin that demonstrated that lightning was a form of electricity gave rise to a new wave of theorizing. Since it was long held that lightning and earthquakes were related phenomena, the idea that earthquakes too were in some way electrical imposed itself. The English clergyman-scientist William Stukeley first presented this idea in 1750 in some articles in the *Philosophical Transactions*, subsequently expanded into a full treatise.[64] A year later, the Italian physician Andrea Bina came up with a similar theory.[65] In the next decades, several other authors presented theories along these lines.

Most of this theorizing appears to have been done in England and Italy. The role of Italian authors is understandable because of the geology of their country. The prominence of Englishmen in the field is a bit unexpected and appears mostly due to the influence of the Royal Society. Interestingly, many of the Italian theoreticians too were in contact with the Royal Society and referred to the works of British experimental philosophers.

All these theories focus on the facts of nature and omit any religious or moral interpretation. That is not to say that these scientists were by definition averse to such an interpretation. William Stukeley was a divine as well as a natural philosopher. At the same time that he proposed his new theory on earthquakes in the *Philosophical Transactions*, he held sermons wherein he explained that the 1750 earthquake was a warning sent by God that people should repent.[66] But by the rules of the new philosophy, such considerations were left out of his scientific work. Natural philosophy as defined by the Royal Society was supposed to focus on the facts of nature and therefore be philosophically and religiously neutral. The Society offered a new "discursive space", a new form of the republic of letters, independent of the demands of teaching and outside the sphere of confessional debate, following rules that had become familiar from travel reports, newspapers, and similar forms.

This new approach to nature did not necessarily result in better research. The Royal Society officially promoted experiments and observations, but it will be clear from the above examples that the professed adherence of English scientists to "experimental philosophy" did not hold them back from speculating wildly. The use of seismoscopes or other instruments was propagated only incidentally, as explained before, and not by leading scientists. True, theories typically had regard to the facts that had been reported as having happened during earthquakes, but British "experimental philosophers" made no systematic attempt to collect such facts or even critically assess the reports. Their evidence is anecdotal or even based on the long-standing written tradition, going back to Aristotle.[67]

Admittedly, since earthquakes are so rare in Britain, English scholars were hardly in a position to do systematic research. Still, it strikes how different their attitude is from the work of some scholars elsewhere. Sturm in Altorf embraced a much more cautious approach. In Switzerland and Italy, some researchers focussed on collecting as many data as possible. In other words, the actual work done was only loosely related to the official ideologies or philosophies. Here again the rise of an increasingly "factual" and "objective" approach to nature appears not dependent on the rise of new philosophical theories.

Notes

1 Cysat (1969) 882, 887.
2 Textor, in Bernoulli (1752) 512. Bernoulli replied that the behaviour was rather due to panic.
3 Hales (1750) 10. Cf. Van de Wetering (1982) 423–424, on sermons.
4 Schechner Genuth (1997); Daston and Park (1998); Jorink (2010).
5 Wendelen (1647). Hartmann (1689). Jorink (2010) argues that the main inspiration for the attack on portents came from philology, not natural philosophy.
6 *True and particular relation* (1748) 101–102.
7 Ruffo (1727, German version) 36. See Mulcahey (2008) 405–407, for comparable discussions about the soil at Port Royal.
8 These circumstances are also narrated by J.F.M.M., *Noticia* (1726) 2–3.
9 Ruffo (1727, Latin version), praefatio by Richter. Richter published his theory on lightning in Richter (1725), see p 25.
10 K. Taylor (2016) 118–119, referring to Paul Aebischer; van Helmont (1648) 99; den Tex (1998) 19. Eckstorm (1620) 175–209 gives a list of twenty *ignes perpetui*, but this includes a burning coal mine.
11 Kircher (1665) 179–182.
12 Ittig (1671). See also Den Tex (1998) 14–18.
13 Cf. Ellenberger (1994) II, 19–21; Willmoth (1987) 35–36. On investigation of volcanoes, see Stokes (1971); Den Tex (1998); Scarth (1999).
14 Mouthaan (2003).
15 Gassendi (1658/1964) 314.
16 J.G. Taylor (1975) 186 states that "the transmission of shocks was a topic of limited interest in the period before 1755." In his view, the theory of subterranean channels begins in 1685 with Robert Boyle. I do not agree with either of these views.
17 Di Somma (1641) 186, 189–191, see also 136–137.

18 Bottone (1718) 55, 99, 127–129. Cf. also Magnati (1688) 123, 415, 418–420.
19 Disputation Jena (1693 Febr.) 45–46.
20 Baglivi (1704) 502–505. Ray (1713) 278–279. Baglivi writes 1694 in stead of 1693.
21 Krüger (1746) 148–150.
22 Nardi (1641). Disputation Helmsted (1656). Disputation Duisburg (1698 July).
23 Scorsone (1993) 353–355.
24 Bottari (1748) 45–48.
25 Taylor (2016) 122, 124.
26 *Unglücks-Chronica* (1692), towards the end.
27 Wettersteint (1673) 5. On this pamphlet, see above, p. 40.
28 Travagini (s.a.) 2. His work is studied in greater detail in Guidoboni (2018).
29 Kochanski (1685). Maybe this was actually the earthquake of Ragusa at April 6.
30 Cirillo (1733) 80.
31 Disputation Utrecht (1747 Febr 22), 11, 67.
32 Hartsoeker (1696) 72.
33 Chr. Huygens (1888–1950) XIX, 311. Petri (1692) 8 also notices a motion from north-north-west to south-south-east.
34 Twynne (1936) 31.
35 Chr. Huygens (1888–1950) XIX, 311.
36 Bernoulli (1752) 502–503 (letter by Textor to Bernoulli, 1737 May 28) 504–511 (a chronological record of the quakes, with times, weather conditions, etc.); 512–513 (Textor's further observations).
37 Ray (1713) 272, 274, 278 (quote).
38 Cirillo (1733) 79–84, quote on p. 81.
39 Bina (1751) 46.
40 Schmidt (1980) 201. Dewey and Byerly (1969) 184–185.
41 Baratta (1895) 6, gives the title and quotes a long passage from the book. See also Dewey and Byerly (1969) 184–185. I have not seen Hautefeuille's original book.
42 Schmidt (1980) 200. Baratta (1895) 7–16, gives an elaborate description of Cavalli's instrument and the observations made with it.
43 Descartes' theory of the earth is explained in the fourth part of his *Principia philosophiae*, from 1644. Descartes (1905) 203–249. Roger (1973) and (1982).
44 E.g. Stay (1747).
45 Lémery (1700). In the accompanying *Histoire de l'Académie Royale des Sciences* for the same year, this article is discussed on pp. 51–52. See also J.G. Taylor (1975) 49–53.
46 Lémery (1700) 103.
47 Lémery (1700) 104.
48 Disputation Gdansk (1728 April 16). See par. 8–9 for the discussion of Lémery's experiment.
49 Bernoulli (1752) 519–520.
50 *Encyclopédie* (1765) XVI, 580–583 (article "tremblemens de terre").
51 Ozanam (1790) 405–406. The paragraph is lacking in the first edition from 1694. Other references: Frézier (1717) 368; *True and particular relation* (1748) 115.
52 Lémery (1700) 103.
53 Lister (1684) 48–50. Taylor (1975) 23–28.
54 Hales (1750). Taylor (1975) 70–75.
55 Woodward (1702) 132–133; see 120 for the water in the earth. J.G. Taylor (1975) 40–43.
56 Disputation Altona (1741 Sept 13).
57 See for an overview of earthquake theories of this period J.G. Taylor (1975).
58 J.G. Taylor (1975) 43–45.
59 Willmoth (1987); Kennedy and Sarjeant (1982); J.G. Taylor (1975) 75–81.
60 Bottari (1733), reprinted as Bottari (1748). The lectures were occasioned by an earthquake of 1727 in Florence and read in 1729 at the Accademia della Crusca.

61 Frézier (1717) 367–369. Taylor (1975) 62–66. Cf. also *True and particular relation* (1748) iv, 102–115.
62 Mezavacca (1692) 263–264. See also Baglivi (1704) 524–525. Interestingly, the copy of *Vera relatione* (1672) in the Newberry Library, Chicago, has on the last blank half page a hand-written draft of a horoscope. The horoscope has no date.
63 *Individual* (1748) par. 2. *True and particular relation* (1748) 134.
64 J.G. Taylor (1975) 91–122. See on the electrical theory also Oeser (1992) 27–28.
65 Bina (1751).
66 Kendrick (1956) 6.
67 See also Willmoth (1987) 36–37.

14

CONFESSIONALIZED NATURAL PHILOSOPHY IN THE AGE OF THE NEW SCIENCES

Both Catholics and Protestants had to find an answer to the challenges of the increase of factual knowledge and the new philosophical ideas that took hold in society. Catholics were initially the most successful in this. The Jesuits in particular were generally informed and well-educated, intent on remaining up to date with the sciences of their time. They too read newspapers and travel reports and in their works, objective data became ever more important. However, that is not to say that they abandoned their religious goals. The teaching of philosophy still had as its objective, apart from the training of the mind, knowledge of all things human and divine. The study of nature stood in the service of that ideal and was defined in religious terms. Jesuits incorporated the new ideas on the earth to the extent that these were useful for their theological interpretations.

Among Protestants, the situation is less clear-cut. Protestant universities were more fragmented than Catholic ones and there was more opportunity for dissent. Whereas the Jesuits managed to keep to a coherent, religiously inspired view of the world, Protestants were divided. Some enthusiastically embraced the new philosophical ideas, but others tenaciously clung to traditional views. The confessional interpretation lost its appeal in many places, but a clear alternative that could command widespread consensus was not immediately in sight. It were mostly laypeople, not theologians or university professors, who suggested new approaches. On the one hand, traditional interpretations in terms of omens and portents lost hold, but on the other, the Bible as a source of knowledge about the natural world was forcefully promoted.

Maybe most interestingly, the clear confessional divides which we noted in earlier chapters started to get blurred. The philosophy of Descartes (himself a Catholic) was first propagated among Protestants, but later gained headway among Catholics as well. Catholic authors like Athanasius Kircher were widely read among Protestants. In the end, both Protestants and Catholics tended towards a view of Nature as an expression of God's providence and his care for all creation, rather than as a theatre of his anger and judgements. This "physico-theology" would become quite dominant in the eighteenth century, but its beginnings can be seen from the middle of the seventeenth century onwards.

Jesuit authors on subterranean fire and the constitution of the earth

Speculation on subterranean fire and the constitution of the earth started well before mechanical philosophy entered the scene, or before the biblical world view lost credibility. Increasingly however, theories on subterranean fire were especially entertained by people who, while convinced of the value of the new factual and experimental approach to knowledge, were unhappy with the new mechanistic tendencies and wanted to keep to some form of a biblical world-view. Subterranean fire could easily be incorporated into those views. As we discussed earlier, in 1643 Franciscus Resta had stated (followed ten years later by Benedictus Mazotta) that the fire that we see erupting in volcanoes was really the fire of Hell, situated at the centre of the earth. This idea would become quite popular in the later seventeenth century.

Especially Jesuit authors would work out a new theory of the earth. They already had a strong tradition in meteorology and were used to incorporate new observations and evidence into their work, though nominally keeping to an Aristotelian framework. Fire got a central place in their physics. The Jesuit Jean Baptiste du Hamel wrote a book "on meteors and fossils" that was published in Paris in 1660. It has the form of a dialogue between Menander (a follower of Descartes), Simplicius, and Theophilus. The latter advocates a modernized Aristotelianism. Du Hamel still paid attention to prodigies. He discussed prodigious rains, arguing that sometimes these were miracles of divine providence, sometimes they were produced by hidden natural causes. Overall, he emphasized the role of the subterranean fire. This is not just the efficient cause of earthquakes, but of nearly everything that is generated within the earth, as well as of many phenomena in the air. Du Hamel too has a long description of siege mines, as earlier in the work of Fromond and Cabeo.[1] He repeated his ideas twenty years later in a general book on philosophy, including meteorology.[2]

The most prominent author on the subterranean fire was the Jesuit polymath Athanasius Kircher. In 1631, Kircher personally witnessed the eruption of Vesuvius, and a few years later he would also experience the great Calabrian earthquake of 1638. These events moved him to devote a major part of his life to the study of the subterranean world and its phenomena. In 1665, he published *Mundus subterraneus* [The subterranean world], a massive work of twelve books in two volumes. In it, he presented a complete philosophy of the earth. Kircher starts with the general constitution of the earth. He continues with the subterranean waters and fires, the origin of mountains and springs, and the four elements, to descend to the properties of minerals and metals, subterranean plants and animals (among them dragons), and all kind of metallurgical, chemical, and mechanical details.

In the earth's constitution, the circulation of fire and water through subterranean conducts takes an essential place. Subterranean fire according to Kircher was elemental fire, although contaminated with various combustible substances. At the time of the creation, God had placed it in the subterranean caves for the protection, conservation, and generation of all things of sublunary nature. All meteors, including clouds, rain, and the like, have their origins in the subterranean fire. All meteors in the air have a counterpart within the earth. Fire and water undergo an everlasting cycle. The fire turns combustible materials into ashes, which, mixed with water, become combustibles for the perennial fire again. The fire could potentially destroy the whole earth, but God has it enclosed within fixed limits, just like the ocean. Still, it is a sign and a preparation of the future conflagration.[3]

Earthquakes and volcanoes take a prominent place in the work. Apart from long narratives in the preface of Kircher's perilous adventures both in 1631 and 1638, the book also contains an elaborate description of Mount Etna, as well as a description by a fellow Jesuit, Franciscus Ricardo, of an eruption at the island of Santorini in 1650. As to earthquakes, Kircher is largely in line with the ideas suggested by Gassendi, Di Somma, and others that they occur above the subterranean fire conducts. For identifying these conducts, Kircher in particular relied on the sounds that the subterranean spirits made both in volcanoes and earthquakes, which were audible on the surface of the earth (he had personally experienced that), and thereby allowed one to follow their course. He thought it indubitable that Calabria by such conducts was connected with Etna, Stromboli, and the other Aeolian islands, but he was not able to establish a connection with Vesuvius. In a later section of the book, on nitre, he made clear that it was not the subterranean fire itself that caused earthquakes, but the sudden kindling of accumulated nitric material.[4]

Although emphasizing that to the extent possible, he had investigated everything with his own eyes, Kircher drew much of his inspiration from ancient sources, among them Platonists. His ideas were well in line not just with the traditional theory of spirits or exhalations, but also with the ancient idea that the earth was a living being. But his emphasis was on the constitution of the earth, of which fire was an essential element, rather than on the question how the fires, spirits, or exhalations were generated or kindled.

Many other Jesuits adopted Kircher's ideas.[5] A few years later, the Jesuit Honoré Fabri published a comprehensive work on physics. Although Fabri keeps within the Aristotelian framework, he no longer follows the traditional organization. There is no separate part on meteorology, but various meteors are discussed at various places. For instance, earthquakes are discussed in the second book, "On *halitus*, and imperfect terrestrial mixtures", of the sixth treatise ("On mixture, and imperfect mixtures"). In his theories too, Fabri not just rehearses earlier authorities. Throughout his argument, he criticizes Fromond. As the main cause of the *halitus* he regards the subterranean fire. All in all, Fabri is clearly more concerned with physics than with theology. He simply ignores supernatural causes or the meaning of phenomena. All effects of earthquakes can be easily explained by natural causes.[6]

The Jesuit Paolo Casati in 1686 wrote a book on fire, wherein he argued that fire was actually the heaviest element in the subterranean world and therefore naturally occupied the central place. He tried to harmonize his Aristotelian world-view with the experiments of Boyle and Toricelli. The central fire was there to torture the damned souls, but also caused the generation of metals in the earth.[7] We find Kircher's impact as late as 1748 in a work on earthquakes written by the versatile author Don Diego de Torres Villaroel, first professor of mathematics at the University of Salamanca. Torres Villaroel also puts Hell at the centre of the earth.[8]

However, Kircher's influence went well beyond the Jesuit order, or even Catholicism. The identification of the subterranean fire with the fire of Hell remained very much a Catholic affair,[9] but Kircher's idea that fire also serves the conservation and maintenance of the world would gain wider approval. In his view, God has placed the fire within the earth for the good of mankind. It is on the one hand a work of God's anger, as it may erupt or shake the earth to punish sinners. Volcanoes are constantly keeping the punishments for our sins before our eyes. But on the other hand, it is a work of God's goodness and enduring providence. Probably for this latter reason, Kircher's work was well received in the Protestant world as well. His book even saw a Dutch translation in 1682. A Swedish disputation in 1703 largely rehearsed Kircher's theory of the earth, but based it all on Cartesian principles.[10]

Protestant authors occasionally hinted at a Christian philosophy in which speculations on the constitution of the earth took a prominent part. The minister Mickel Escholt, from Christiana (present day Oslo), published a long history of the earthquake in southern Norway in 1657. He referred to a number of learned theories. Although the main thrust of the text is theological, he paid considerable attention to the constitution of the earth and the origin of metals. In his view, the earth is hollow, filled with vents, waters, and fire. His description is a mixture of theology, established philosophy, and some speculation akin to the lay philosophers to be discussed later.[11]

New tendencies at protestant universities

In the Protestant world, the confessional view had been forcefully propagated by the universities. However, unlike in the Catholic world, there was no central authority that enforced a common philosophy. Debates at universities, as apparent from disputations, appear to have been more lively than textbooks suggest. Moreover, from the second half of the century onwards, professors got more possibilities to venture their ideas. Earlier disputations were in most cases written on some standard *quaestio*, mostly taken from Aristotle's works. These standard questions required standard answers. In the second half of the seventeenth century, disputations became more varied and addressed all kinds of topics. Traditional meteorological disputations on the other hand all but disappeared.

All professors of philosophy had to come to terms with newly emerging philosophical ideas, above all Cartesianism. Cartesianism therefore frequently turns up in academic disputations, either to be praised or refuted. This concerned not just the mechanical principles of natural things, but also the relation of God and nature. Descartes allowed only natural causes and refused to include discussing God's intentions in natural philosophy. All phenomena, common and strange, are brought about by the laws of nature, which are universal and immutable. Visible reality therefore should be explained based on these laws of nature alone, without any reference to God's plan or intentions. This was a fundamental break with confessionalized philosophy, but also with most other traditional philosophies, for whom the investigation of nature served wider aims of understanding religious and metaphysical truths.

The author of a disputation on earthquakes at the Dutch university of Franeker in 1677 has clearly undergone the influence of the "new philosophy". Although in the corollaries, he rejects Descartes' theories that animals are just machines, he also claims that their *forma* was not a *spiritus*, but the disposition of their parts. In the main text, he gives only a brief nod to the religious aspect. He admits that many people feel that earthquakes are direct acts of God, without natural causes, but he will just talk about the natural causes, and the species, magnitude and duration of earthquakes. Indeed, the disputation is silent about any religious significance or biblical connotations. After a long discussion of the various causes that have been proposed for earthquakes, the author concludes that they are not caused by vapours, but by fumes consisting of fiery particles. Unoriginally, he sees an analogy with gunpowder. All phenomena that are found in descriptions of earthquakes can be explained in this way. When a large part of the earth (or even the whole earth) quakes, this is because the caverns where the fire is burning are very deep. "For as the cavern is closer to the centre of the earth, a larger mass of the earth will be agitated."[12] So, the presumption that an earthquake that shook the whole earth (as at the Crucifixion) should be miraculous, is implicitly refuted.

Some disputations defend a Cartesian point of view. At Bremen, a disputation defended under the presidency of Johann Eberhard Schweling in 1684 gave a history of earthquakes and a physical explanation. Schweling was a follower of Descartes and his explanation consists mainly in a rehearsal of Descartes' theory of the earth. However, he is not leaving religion alone. In his history, he does include biblical quakes and also the quake under Julian the Apostate. In a last section, the author made a distinction between physical and miraculous earthquakes. Among the latter were counted all biblical quakes, especially those during Christ's death and resurrection, a few from classical history and, in the opinion of the author, the earthquake of *Droborniram* (Dover?). All other quakes are merely physical, as are most other events commonly regarded as portentous. The author definitely rejects prodigies – actually, his very title indicates that he will explain all earthquakes physically. That includes the so-called miracles. As a

Cartesian, the author stresses that even in miracles, God does not act against the laws of nature.[13]

The author of a disputation at Uppsala in Sweden from 1686 stated that he would skip the preternatural earthquakes of which the Bible speaks. He largely followed the idea of natural fire and subterranean canals. Referring to Descartes' "Principles of philosophy", he argued that earthquakes and volcanoes had the same cause. He explicitly rejected Pliny's idea that earthquakes announce evil: "Since an earthquake necessarily happens by natural causes, I do not see in what way it can predict the future."[14]

However, other professors resisted the tendency towards natural explanations and emphasized the prodigious nature of many phenomena. In 1664 at Wittenberg, the professor Sebastian Kirchmajer presided over a disputation on images in the sky. Kirchmajer rejected all natural explanations of such phenomena. He maintained that such *ostensa* were

> nothing else than vapours or exhalations, which have been given in some region of the air the shape of real things by some non-natural cause, in order to announce some extraordinary event. And this is said not as it pleases, but as it is allowed.

Kirchmajer left open the question whether God worked directly or indirectly, that is by means of angels or demons. Also he was not sure whether God used vapours that were already present or whether he created them specially for the occasion, although the first seemed more probable.[15]

Johann Michael Schwimmer, professor at the Athenaeum at Rudolfstad, held a disputation on the 1690 earthquake of Villach as it had been felt in Thuringia. Schwimmer was not interested in the physical causes of the phenomenon, or in an exact description. Instead, he focussed on its prodigious nature. He included a long discussion of the nature of prodigies, wholly in the spirit of the earlier discourses at Wittenberg. In the end, he concluded that the earthquake had been a divine miracle – a *prodigium verum et divinum*. God acted as its efficient cause, both general and particular (*causa particularis et principalis*).[16]

Some philosophical disputations, especially at Lutheran universities, allotted a central place to the Bible. By the second half of the seventeenth century, the earthquake that accompanied Christ's death on the cross became a set topic for philosophical disputations. So, at Jena in 1672, under the presidency of Caspar Posner, physician and philosopher, one Petrus Brand wrote and defended a thesis (*dissertatio physica*) on earthquakes, "particularly those which accompanied the death and resurrection of Christ". At Helmstedt in 1673, the physician Valentin Heinrich Vogler dealt with the same earthquake in his "Physiology of the history of Christ's passion". In 1683, Christopher Weisse wrote and defended a disputation at Jena under the professor of logic and metaphysics, Johann Andreas Schmidt, on the earthquake at Christ's passion. Johann Fridericus Köber wrote a dissertation in 1683 on the earthquake and the solar eclipse at Christ's death.[17]

The most pressing question in these disputations was whether the earthquake had been effected by natural or supernatural causes. The authors invariably concluded to the latter. In some cases, they tried to connect the Biblical story with known events from ancient history. Köber identified the earthquake at Christ's passion with one in Bithynia, mentioned by ancient historians, and concluded that it had been world-wide (and thus miraculous).[18] However, his view was not universally accepted. Weisse, writing in the same year, doubted whether the earthquake in Bithynia was the same as the one during the Passion, although he did not doubt that the earthquake was universal; after all, the Bible says that "the earth" (ἡ γῆ) shook. Strauchius likewise denied the identity of these earthquakes. Most authors agreed that the earthquake was universal. Posner however, although not doubting that the earthquake was miraculous, left the question of its universality undecided.[19]

The most common argument was based on natural philosophy. Posner discussed a large numbers of authors and gave a detailed explanation of the natural causes of earthquakes, concluding that the earthquakes at the Crucifixion and the Resurrection were not of such a cause and origin. They therefore had a divine, supernatural cause. Weisse's disputation has a similar structure. He first gave a detailed account of what was known about earthquakes in philosophy. From these, he argued that the specifics of the earthquake at Christ's passion: its force, its short duration, the various signs that accompanied it, could not be explained naturally. The earthquake must be supernatural. Vogler too, after discussing the natural causes of earthquakes, concluded that the earthquake at the Crucifixion "was without contradiction simply miraculous". The same holds true, he claimed, for the concussion at Horeb (or Sinai) at the pronunciation of the Ten Commandments and for the future earthquakes at the end of times.[20]

By emphasizing the supernatural character of biblical miracles, in this case, the earthquake at Christ's Passion, these authors appear to have reacted to contemporary developments in philosophy. The miraculous had become a problematic and worrying concept in natural philosophy, and the biblical miracles played a key role in this debate. The status of such miracles would remain a much debated subject for decades to come.

It would seem that the view of earthquakes as prodigies lost its hold at Protestant universities not because of a conscious effort. It gradually eroded. Professors were contemplating the various theories that were around. Since prodigies were no longer seen as essential in the social and political order, as they had been in the sixteenth century, there was no reason any longer to prefer this interpretation above its competitors, old or new, and it just faded away.

Protestant lay philosophers

Debates at universities followed the various tendencies, but they do not appear to have themselves initiated any major new insights. These mostly came from outside the universities. Many laypeople wondered about the religious meaning and consequences of the new physical discoveries and principles. Some were trained physicians or other intellectuals, but the question was deemed relevant not just by them. As a direct consequence of the new media, many more people than before could join the intellectual debates. One no longer needed to be trained in Latin and scholastic methods or have access to expensive scholarly works on history and philosophy. Factual information was all they needed and was easily accessible via works in the vernacular.

Under these circumstances, a group of people emerged that we may call lay philosophers. These people could be schooled or unschooled in traditional learning, but they had lost faith in scholastic philosophy. Consequently, they looked elsewhere. They were avid readers of all kind of literature: natural-philosophical, medical, but also historical and biblical. From these, they picked elements to help them understand the world. This led to a, for modern eyes, often somewhat hybrid character of their ideas. On the one hand, we find in their works the same objective treatment as we saw in travel accounts and newspaper reports. Often, they had some inkling of the importance of a mathematical or empirical approach. On the other hand, they certainly did not do away with religion. The Bible is often one of their main sources. But then, they mine the Bible mostly for information on the physical world. Very few of these people were brilliant thinkers; most of them were at the fringe of intellectual life, and some of them were clearly lunatics. Still, even if their work left little trace in scientific theory, the very presence of such ideas attests to the changing understanding of the world.

One such lay philosopher was Alexander Achilles, a Prussian army officer. Not much is known about him, but apparently he had travelled widely. In 1666, he published a book wherein he claimed to explain the causes of earthquakes and related phenomena; a second part speaks of the tides. Achilles was well aware of the progress of the investigation of nature achieved in his time, which had made that hardly anything on the earth or at sea was left to discover, "except for the internal and lower world and what is hidden therein, [which] has been reserved for me to bring to the light, nearly as an untimely and unlucky birth".[21] As is more often the case with amateurs, Achilles emphasizes his practical knowledge as opposed to bookish learning. Even on the title page, it is not said that he has written the book, but that he has "experienced it himself".[22]

He touches on a variety of phenomena. His arguments consist mainly in drawing analogies with familiar things. Earthquakes are caused by terrestrial exhalations, when they are impeded from emerging by frost or rain. In cities, the many walls and other constructions too block their way, so that they need more force to escape. That is why earthquakes in major cities tend to be stronger. The digging of holes is an effective remedy against earthquakes. All this sounds familiar, of course. Achilles diverges from Aristotelian theory mainly by denying a relation of these exhalations with the wind and that they move in subterranean caves. Instead, he feels that all the minerals in the earth generate specific vapours (*Dämpffe und Gewittere*; in the second part, *Gewittere* are explained as "wet fires").[23] These vapours are responsible for all kinds of phenomena − tornadoes, blood rains, images in the air, plague, and medical properties of plants. Interesting is that he combines this with some detailed ideas about the inner structure of the earth.

At the same time, his work is impregnated with biblical thinking. As he states in the dedication, he will treat the matter "as far as God and experience has granted me". At several instances, he refers to the Creation. For example, he argues that the metals must have been created at the divine *Fiat* (that is, they have not grown gradually in place), because otherwise, the earth for many years would have remained hollow. He upholds the analogy between microcosmos (the human body) and macrocosmos (the body of the earth). Under the earth, there is enough fuel and cooling to torture the damned in eternity.[24]

In his dedication, he rather elaborately defends himself against the accusation that he improperly tries to grasp by reason things that are above nature. Achilles replies that God has created the whole world according to nature and keeps it in its proper order. Only the prime matter (*prima materia confusa*) is created out of nothing. If from this matter God had wanted to create something against nature, he could have done so instantaneously and would not have used six days. He wanted to follow the order of nature, in which everything will remain till the end. The natural things are there for man to investigate them. Nothing is so far, high, or deep that it cannot be attained by wisdom and virtue.

A somewhat similar figure is Jakob Ziegler, a physician from Zürich. When in 1674 Switzerland was struck by an earthquake, a publisher asked him to write a short tract on the natural causes of earthquakes. Apparently, in local circles, he had already gained some notoriety as a man with bold ideas. A local minister, "kindly requested", added a poem to the work.[25]

Ziegler's speculations are based partly upon the familiar idea of vapours and exhalations as causes of earthquakes; partly on analogies with some familiar phenomena; partly, it would seem, on biblical elements. He starts with a description of the earth. In his view, the earth is composed of various layers. The top layer consists of black, fertile soil. Below, there is a layer of infertile, solid ground that is impermeable to water or air. This layer separates the rainwater above from the source water below. Under the source water there is still another layer of earth, and below that, there is again water. This water is the water of the sea. Under the sea water, the earth finally stretches to the centre.

All the countries and islands therefore are floating upon the sea; they are founded upon the waters, as stated in Psalm 24:2 and 136:6. Now, the upper and lower surfaces of the lands are parallel. The upper surface is convex (since the earth is a sphere), so the lower is concave. Under these concave surfaces, there are vapours and air that carry the islands, just like a ship is kept afloat by the air inside. At another point, he compares them to a bowl put upside down into the water. (Of course, there is a logical flaw here: the surface is concave, but since all its points are equidistant to the centre of the earth, it still composes a horizontal plane, which cannot keep air trapped.) The inner core of the earth, because of the adjacent sea water, has its own mixture. From it rise air bubbles, just as one observes in stagnant lakes. When either the production of bubbles is too strong, or the pores in the earth are obstructed by rainy weather, too many vapours assemble under the continents or islands and in the end, they force a way out. Thereby, they lift the earth; once the vapours have escaped, the earth has to settle again, which cannot happen without violent quaking.

As a medical doctor, Ziegler must have been well grounded in learned literature, but his theories do not show many traces of that. The idea of vapours and exhalations is here transformed into a simple mechanical model, a bowl put upside down into the water. The idea that the earth consists of layers is partly based on observations, partly it may owe something from authors like Descartes. The idea that the lands are floating upon the sea appears to be taken from the Bible, but it is corroborated with some geographical literature on Scotland.

These writings are remarkable both for what they do contain and for what they do not contain. These lay philosophers emphasize experience and empiricism, but without abandoning religion. Biblical elements are treated simply as other empirical data. However, there is hardly any interest any more in the prodigious nature or the meaning of earthquakes. The authors are mostly concerned with understanding earthquakes from what they know (or think they know) about the constitution of the earth. The prodigious interpretation is losing its grip, even though religion itself remains of prime importance.

The Deluge, the Bible, and the rise of physico-theology

By the end of the seventeenth century, confessionalized philosophy was on the wane. However, the rejection of prodigies did not imply a complete secularization. Hand in hand with the dismissal of the idea of an angry and vengeful God, the idea of a benevolent and providential God gained prominence. Kircher's suggestion that fire, earthquakes, and volcanoes served a purpose in the economy of nature and the providential ordering of the world was widely accepted and became a kind of general template for understanding the natural world generally. These tendencies would eventually result in a powerful lay philosophy, mostly but not exclusively in the Protestant world, which is known as physico-theology. Debates on earthquakes are part of this wider movement.[26]

An interesting case is the history of the Sicilian earthquake of 1693 by Domenico Bottone. This work was published only in 1718, when orthodoxy was on the wane. The book of over a hundred pages relates in much detail the various physical phenomena that accompanied the earthquake (not just movements in the earth, but also meteors and the weather). As common in such histories, the book passes in review the various cities that were struck and relates the damage and the victims, but there are also various scientific excursions. Among other things, Bottone discusses whether putrefying bodies cause plague and investigates the cause of earthquakes, in particular the role of subterranean fire and of Mount Etna. There are many references to Kircher.

The book is critical of prodigies. Strange effects are explained from simple natural causes. Bottone reported that at the earthquake, a stream near Messina had turned black and gave a sulphurous, bituminous smell. On investigating the case, he had found that the stream obtained its colour at a hill that contained much coal. The quake had shaken the hill and the coal had got mixed with the water. It occasioned him to speculate on the role of coal in earthquakes.[27]

However, God is not completely ignored. Actually, Bottone dedicated his book to the "Most exalted powers of the heavens and the divine protectors of the terraquaeous globe" [*Serenissimis Caelorum Potestatibus Terraquei globi Tutelaribus Divis honor*]. But there is no reference to God's anger. Rather, Bottone emphasizes the providential ordering of the world. His theory of subterranean conducts led him to the view that sulphurous springs served to rid the earth of sulphur and saltpetre, which otherwise would accumulate dangerously underground. The same was true for Etna: God had given it as a kind of safety valve. Bottone ridiculed the fear of the people of Sicily that the fire of Etna in the end would destroy the whole island, a fear that was (as was common) incited by some self-proclaimed prophets.

> As we do not dare to reach to divination, we will descend to human things and to experimental philosophy, according to which we feel that it is stupid to accuse Etna of a crime and to attribute to it the disaster of Sicily.[28]

An addition to the text, on an earthquake in 1717, also mentions God, but as a saviour, not an avenger: through God's protection, the quake made no victims. Referring to Augustine and to Jesaja 41, Bottone specified that the future is hidden from man. Some people used the disasters to impress the common man with the need for piety, but a Christian does not need such warnings. One should have God's judgement always on one's mind.[29]

Similar views took hold in the Protestant world as well. A disputation on earthquakes at Utrecht from 1692 is still almost completely Aristotelian. The author even makes a side remark on Copernicus' theory, which he calls against scripture. But for the rest, the disputation is almost completely philosophical. Only natural causes are discussed and the role of earthquakes as instruments of God's anger is ignored. To the contrary, earthquakes have their legitimate place in the world's order. By making vapours escape from the earth, they prevent the corruption of the terrestrial world. Rather than as a theatre of sin and punishment, the author describes the world as a perfect order.[30]

Such optimism would become quite prominent in the eighteenth century, until Voltaire ridiculed it in his poem on the Lisbon earthquake of 1755. Another Utrecht disputation on earthquakes was defended in 1747. The author is rather eclectic and refers to many ancient and modern authors. The disputation is very elaborate, but only natural causes are discussed and the notion of punishment is completely absent. On the contrary, the work starts with a five-page preface on divine providence and ends with an epilogue on the use of earthquakes. The subterranean fires generate and perfect metal ores. They give rise to healing springs. The earth is rendered fertile. Everything is dependent upon God and even such harmful phenomena as earthquakes can be put to use.[31]

However, although the concept of prodigies lost credibility, Protestants increasingly referred to the Bible in their natural philosophy. This was a tendency we noticed before in academic disputations and among lay philosophers. By the end of the century, the Bible became a constituent element of Protestant natural philosophy, propagated not so much by university professors, but by natural investigators eager to give a religious legitimation to their research. In particular, the debate on the biblical Deluge gave the tendency to study earthquakes with a focus on biblical history a huge momentum.

The debate was set in motion by the English divine, Thomas Burnet. In 1681–1689, he published his book *Telluris theoria sacra*, which within a few years was translated into English (*The sacred theory of the earth*), Dutch, and German. In this book, Burnet explained the biblical Deluge as having occurred by natural causes, a consequence of the constitution of the earth as originally created by God. The book was highly controversial, not the least because Burnet, in order to argue that his theory was in agreement with Scripture, explained some of the biblical sentences in a rather unorthodox way. Following the publication, many people started discussing the constitution of the earth in relation to the biblical Deluge.[32]

In these debates, the principle of adhering to the Biblical truth was combined with a factual, empirical approach. Much of the work of course remained speculative, but the debate also gave rise to interesting research on topics like fossils and stratigraphy.[33] Burnet himself had not much to say on earthquakes, but in the course of the debates, these unavoidably turned up. A major issue was the question how the mountains, oceans, and other structures on the earth's surface had been formed. Burnet theorized that the earth had originally been a perfectly smooth sphere. Irregularities resulted from the collapse of the earth's crust at the time of the Deluge. As such, mountains were deformities and a visible proof of the fallen state of our present world. Later authors came up with alternative theories and many came to see here a role for earthquakes.

True, not all of these authors were concerned with defending the biblical truth. A few years after Burnet had published his book, Nicolaas Hartsoeker published *Principes de physique*. It includes a long section on the history of the earth. Although he does not refer to Burnet or the Deluge, it seems clear that this debate is on his mind. Hartsoeker argued that the earth consists of several beds of matter, with a very subtle matter in the centre. Bodies in the earth formed vaults over each other, from which originated mountains, valleys, and cavities. This was demonstrated in earthquakes. Without mentioning Lémery, Hartsoeker subscribed to his idea that earthquakes were caused by a mixture of sulphur and iron. The presence of such a mixture in a subterranean cave that also contains compressed air, will cause an expansion of the air. This will somewhat lift and shake the vault of the cave, sometimes even make it burst and collapse.[34] Hartsoeker referred to several ancient histories, but then came to speak of the earthquake of 1692 that he had witnessed at Amsterdam. The fact that this earthquake was felt in the Low Countries, the British Isles, France, Germany, and Denmark, seemed to indicate that all these regions are situated on one big vault.[35] Hartsoeker can be read as giving a physical explanation of some biblical elements, such as deluges and universal earthquakes, but he carefully avoids to mention the Bible.

Most of the debates however focussed on the authority of the Bible. The English naturalist and divine John Ray, a respected member of the Royal Society, took position against Burnet in his *Three physico-theological discourses*. This book discusses the possible causes of the Creation, the Deluge, and the final conflagration. Ray used scientific methods to get closer to the biblical truth. He discussed the changes that had been made in the earth's surface by the general Deluge or otherwise. Against Burnet, Ray suggested that mountains could have been generated by earthquakes. In a traditional way, he felt that earthquakes had their origin in combustible subterranean vapours that caught fire. Of what kind these vapours were, and how they spontaneously could catch fire, he was not quite sure.

Ray discussed the natural causes of earthquakes at length; his argument is built on many empirical observations. In the end, however, his motivation is religious. He did not doubt that earthquakes are sent by God to punish us. In his book, he included a description of the 1692 earthquake of Jamaica, as well as of the earthquake that had been felt later that year in England. Ray definitely felt that the earthquake of Jamaica was not just caused by natural means. As the people of the colony of Port Royal, which had been hardest hit, had been so godless, the earthquake might rightly be kept for a judgement of God, of the kind as had earlier happened at the time of the Deluge, or with the towns of Sodom and Gomorra.

> For God does not stand by as an idle and unconcerned Spectator, and suffer Things to run at Random, but his Providence many times interposes, and stops the usual Course and Current of Natural Causes. So, though the Instruments and Materials wherewith this Devastation in *Jamaica* was made, as a subterraneous Fire and inflamable Materials, were before in the Earth, yet that they should at this time break forth and work, when there was such an Inundation of Wickedness there, and particularly and especially at Port-Royal, this we may confidently say, was the Finger of God, and effected perchance by the Ministry of an Angel.[36]

Ray combined this traditional religious outlook with a rigorous scientific method. His book has remained known for the empirical research it contains on fossils, which he studied as relics of the Deluge, and geological phenomena. Although he did describe earthquakes as instruments of God's anger over our sins, at the same time, he emphasized God's providence in ordering the natural world. Herein, he demonstrated God's wisdom and benevolence in arranging everything for the benefit of man.

The Swiss physician and naturalist Johann Jakob Scheuchzer is probably the most important representative of the new empiricism in the investigation of earthquakes. He studied under Sturm at Altorf, who impregnated him with his empirical approach. After taking his doctorate in the Dutch Republic, he returned to his native Switzerland where he settled as a physician at Zürich. He was well informed of the current state of science and was in close contact with the international learned world, in particular the Royal Society of London. Still, he regarded himself as a student of Sturm and his eclectic philosophy.

Religion keeps an important place in Scheuchzer's thinking. He wanted to harmonize the new science with the demands of religion, not always to the satisfaction of the orthodox. To this purpose, he wrote extensively on the physics of the Bible. His aim was to demonstrate the biblical truth. In this context, he paid considerable attention to the Deluge. Like Ray, he collected and studied fossils, which he too saw as relics of the Deluge. Earthquakes too got their meaning from the Bible. It goes too far to say that it was just because of his interest in the Deluge that Scheuchzer studied earthquakes, as he was interested in many aspects of natural history, but it certainly informed his research.

Scheuchzer wrote major works on the natural history of Switzerland and the Alps. Apart from mountains, minerals, and fossils, he also took an interest in earthquakes and aimed to incorporate them into his natural history. In his theories, he was not innovative. He regarded earthquakes as caused by the kindling of sulphurous and nitrous vapours in a subterranean system of caverns and canals.[37] Methodologically, however, he belongs to a new period. Scheuchzer was much more empirically minded, and therefore cautious, than his predecessors.[38] In his *Hydrographia Helvetica*, he considered hot springs. He felt it credible that in Italy, sulphur was the cause of such springs, but this mineral was lacking in Switzerland. From his observations, he concluded that hot springs occurred where the heat that was generated in the earth could not well escape, but

> whether this internal heat of the earth derives from a blazing fire, or from a specific glow or boiling, as some modern philosophers deduce from the motion of the earth, or from still other causes, I do not want to decide at this moment.[39]

As an empiricist, Scheuchzer attached much value to the compilation of data, a tendency that became only stronger later in his career. He collected unpublished eyewitness reports and other sources and included some of these in his books. Some of these were already pretty old in his time, like a letter on the disaster of Yvorne in 1584 by the physician Johann Rudolf Bullinger from Bern, who had visited the place a few weeks after the event, and a Latin poem made on the occasion of the earthquake at Unterwalden of 1601.[40] As he did for other phenomena, he compiled extensive chronicles of earthquakes and landslides. These catalogues are a work of compilation rather than critical history. Still, their aim is no longer to find a theological pattern, but rather to uncover the order and predictability of nature. Earlier earthquake chronicles had mostly been a list of earthquakes in world history. Scheuchzer limited himself to his native Switzerland.

Scheuchzer's earthquake research was initiated by an earthquake in the canton Glarus on 22 May / 3 June 1705. This led him to compile a catalogue of earthquakes at Glarus covering the period 1654–1703. Of course, he had to rely on existing records, in particular a recent description of Glarus by J.H. Tschudi. Later, Scheuchzer expanded his list to a catalogue that covered the whole of Switzerland for the period 849–1726.[41] J.G. Sulzer, who in 1746 oversaw a posthumous edition of Scheuchzer's work, praised Scheuchzer for this work which, he claimed, when it was first published had appeared "very repugnant" (*ganz eckelhaft*) to many people who were unfamiliar with the methods of natural investigation.[42]

Scheuchzer noted carefully the various phenomena that accompanied the earthquakes in Glarus: a shock in the soil under one's feet; a shaking of the earth which made move and fall down big rocks "according to the report of the farmhands on the alps"; and a humming in the air. He pointed out that all these phenomena were in agreement with the current theories of earthquakes. He gave a number of observations that indicated that these earthquakes were not caused by subterranean wind, but by some burning fire: the falling dry of sources, the smell of sulphur, unusually warm weather and early disappearance of snow, and cases of smoke rising from the earth. He also argued for the existence of hollow spaces under the mountains.[43]

Scheuchzer corresponded with people in other places who informed him on local earthquakes. Johann Bernoulli, professor of mathematics at Basel, informed him of an earthquake felt in that city on 9 February 1711. Bernoulli's report was not just concerned with the quake itself, but with other circumstances as well. Bernoulli noted that before the earthquake there was a strong wind, whereas after the event, a strong south wind (*Fön*) was blowing, that melted the snow and made the river swell. This *Fön* lasted only a few hours, after which the cold returned, with much snowfall. "These circumstances have given to the learned professor Bernoulli at Basel the thought, that this warm wind somewhere has erupted from the subterranean caves, with such violence that it shook the earth; to which idea I completely subscribe."[44] Scheuchzer felt that his own observations corroborated this idea. Such an unleashing of a warm wind would have brought a change of the air in the rest of Switzerland as well. Indeed, he noted that at Zürich, the barometer had fallen unusually low at the time of the Basel earthquake.

In the 1720s, Scheuchzer turned to a more systematic collecting of data on earthquakes that happened in his own time. He wrote several case histories, based on eyewitness accounts which he actively collected with the help of a wide net of correspondents. He was not satisfied with spontaneous records, but wanted specific information on the date, place, size of the affected area, damages, accompanying phenomena, and reactions – an early and primitive form of standardization. He attempted to relate specific earthquakes to the local constitution of the soil, in particular the occurrence of sulphur and nitre, or to the presumed existence of large subterraneous caves which might collapse. He also paid attention to the emergence or disappearance of springs, which he explained from fires or watery vapours under the earth. Scheuchzer also made a distinction between earthquakes and non-seismic events (*Berg-Fälle*), as those that had happened at Yvorne and Piuro.[45]

Scheuchzer's religious ideas are quite distinct from those of the Reformation era when it comes to prodigies. He was very critical about signs and portents and tended to disqualify most of them as superstitions. The fiery meteors that were sometimes seen were simple natural phenomena. People regarded them as wonders because they were uncommon, but if rainbows would be uncommon, people would regard these as portents as well.[46] Prodigious rains in most cases could be explained as mere natural phenomena. "I am not so credible in this case, and strongly doubt whether it has ever rained blood in the world."[47] In his "Description of the natural histories of Switzerland", he discussed a large number of reports, taken from chronicles and other places, of fiery meteors and images in the air: flying dragons, burning lances, "fiery men" (=will of a'wisps), and so on. All these are regarded as natural phenomena. On the fiery sign that was allegedly seen before the battle of Dornegg in 1499, he commented: "This way to turn natural histories into prophetic wonders, which derives from ancient pagan times, still adheres to us learned and unlearned. So hard is it to get rid of superstitions that one has imbibed with one's mother's milk."[48]

Still, Scheuchzer is cautious not to deny any higher meaning at all. He justified the inclusion of a number of strange phenomena in his book with the words: "Who is a lover of a high supernatural and hidden philosophy, may be served with the following wonders [*Wunderzeichen*]…" However, he limited himself to a description and refused to interpret them: "I confess my ignorance in these exalted mysteries."[49]

When it comes to the meaning of earthquakes, Scheuchzer was ambivalent. On the one hand, one of his main tenets was that God had organized the creation in a most wise and perfect way. Earthquakes too had an important role in the overall economy of nature and therefore showed God's wise contrivance in the Creation. Still, Scheuchzer was unwilling to reject all notion of divine punishment. He praised God's providence in the Swiss country. Nature has been organized in a way to provide for the needs of the inhabitants. On the other hand, "exactly this so praised omnipotence of the Creator is also capable, given our mounting sins and abuse of our country's benefices, of punishing us at any moment according to its justice".[50]

The same ambivalence can be seen in his chapter on *Berg-Fälle*, collapses of mountains. He discussed the cases of Yvorne and Piuro in detail. He explained that it was not strange that mountains now and then collapsed. It was much stranger that this happened so rarely. Mountains according to him are hollow and are constantly being eroded by wind, water, earthquakes, and so on. "Here one discerns an amazing providence of the all-good creator. He had to arrange this big construction in such a way that it was both wacky and permanent, hopefully until the end of time." Not only Switzerland, but the whole of Europe drew great profit from these mountains. "Still, God warns and punishes now and then by means of *Bergfälle* that happen here and there…"[51]

Earthquakes are natural, but they are also signs that the Day of Judgement is near. The fact that we are uncertain about the causes of earthquakes, should be reason for us to speak of them only in a holy fear and as warnings to lead a better life. Scheuchzer closed the first volume of his "Natural history of Switzerland" with the words:

> The more we have great cause to atone us with our God, for our Swiss earth-vault, burdened with mountains, has already suffered many quakes; and we can be confident, both from these natural causes and from the moral causes, I mean the rising sins and evildoings, that the *terminus fatalis* is close, as God following his strict justice and inscrutable wisdom can submerge the pillar of our country, and have us devoured by the earth itself.[52]

Ray and Scheuchzer were learned scholars, but no university professors. They wrote partly in the vernacular and their works had also an edifying tenor. Both were very influential authors who largely set the tone for eighteenth-century physico-theology. Still, they made the move to a providential, benevolent God only partially. Though sceptical of portents, they kept to the view of a vengeful God who uses natural disasters to punish human sin. Most noticeable is their use of biblical data in a modern-empirical framework.

Notes

1 Du Hamel (1660) 5–11, 10 (siege mines), 54–55 (prodigies).
2 [Du Hamel] (1682) II, 413–416 (earthquakes, subterranean fire), 454–455 (prodigious rains).
3 Kircher (1665) I, 169, 174, 183, 189, 219–225. I also used the Dutch translation: Kircher (1682).
4 Kircher (1665) I, 186–190 (Etna), 221–222 (conducts); VI, 307 (nitre).
5 See especially Capel (1980).
6 Fabri (1669–1671) III, 274–287. Proposition 29 (284–286) states: "Effectus, qui vulgò terrae motui tribui solent, ad suas causes physicas facilè reducuntur."
7 Casati (1688).
8 Torres Villaroel (1748). See on him Capel (1980) 28–34.
9 An exception is Alexander Achilles, on whom below.
10 Kircher (1682). Disputation Uppsala (1703 Oct. 4).
11 Escholt (1663).
12 Disputation Franeker (1677) 12, thesis 16.
13 Disputation Bremen (1684 May 3) aphorismi 13–17, 49–53.
14 Disputation Uppsala (1686 May 17) section 10. The reference to Descartes is to book IV section 78.
15 Disputation Wittenberg (1664).
16 Disputation Rudolfstad (1691) par. 16.
17 Disputation Jena (1672 May); Vogler (1673); Disputation Jena (1683 Nov. 16); Köber (1683).
18 Köber (1683), passim. The chronicle by Johannes Nauclerus stated that at the time of the Passion Bithynia and especially Nicea, although far from Jerusalem, was struck by an earthquake: Nauclerus (1614) 427. Many Catholic authors held the same view, cf. Tirinus (1632) III, 82.

19 Disputation Jena (1683 Nov. 16) sectio I; disputation Wittenberg (1608) th. 4; disputation Jena (1672) 36–39.
20 Disputation Jena (1672) 41–45; disputation Jena (1683 Nov. 16) sectio III; Vogler (1673) 41.
21 Achilles (1666) preface to the reader. The book has no page numbers.
22 Achilles (1666): "welche selbst erfahren hat…". See also section 20, wherein Achilles emphasizes that what he writes on the medical qualities of minerals etc. has not been taken from books, but from experience.
23 *Gewittere* literally means thunderstorms, but Achilles is likely thinking of *Witterungen*, mineral fumes.
24 Achilles (1666) section 11.
25 Ziegler (1674).
26 The exact definition, demarcation etc. of physico-theology remain a matter of debate. For a state of the art overview, see the recent volume by Blair and von Greyerz (2020).
27 Bottone (1718) 13–14.
28 Bottone (1718) 53, see also 53–54, 130. Cf. Condorelli (2013) 161.
29 Bottone (1718) 52–53, 125.
30 Disputation Utrecht (1692 March 29). See esp. pages 3–4 (theses 1–2) and 14–15 (thesis 20).
31 Disputation Utrecht (1747) 1–5, 67–68.
32 Barnett (2019) 89–128 and passim.
33 Rudwick (1976).
34 Hartsoeker (1696) 209–211.
35 Hartsoeker (1696) 68–73.
36 Ray (1713) 271.
37 Scheuchzer (1703) 179–185. Scheuchzer (1746) I, 178–180. Scheuchzer (1706–1708) I, 117.
38 On Scheuchzer's research on earthquakes, see in particular Gisler (2000).
39 Scheuchzer (1717) 325.
40 Scheuchzer (1716) 128–132. Scheuchzer (1718) 85–87. The poem is by Johann Huldrich Grobius and addressed to Johann Jakob Frisius.
41 See Scheuchzer (1706–1708) I, 118–120, and (1746) I, 180–184 for the list from Glarus; Scheuchzer (1706–1708) I, 123–128, and (1746) I, 186–194 for the list from Switzerland. Scheuchzer (1746) II, 360–367 gives additions to this list for the period 1720–1726.
42 Sulzer, annotation in Scheuchzer (1746) 179.
43 Scheuchzer (1706–1708) I, 121–123.
44 Scheuchzer (1716) 94.
45 Gisler (2000).
46 Scheuchzer (1718) 46–47.
47 Scheuchzer (1718) 17.
48 Scheuchzer (1706–1708) II, 50.
49 Scheuchzer (1746) II, 80.
50 Scheuchzer (1746) II, 79.
51 Scheuchzer (1716) 127. The whole chapter is on 127–144.
52 Scheuchzer (1746) I, 194–195.

15

EARTHQUAKES IN THE RELIGIOUS DISCOURSE OF THE LATE SEVENTEENTH CENTURY

The various tendencies in the seventeenth century did not merge into a new synthesis. They existed side by side, or rather, they increasingly grew apart. As the new empiricism of the seventeenth century tended to ignore religious and moral dimensions, so the religious message in the various Churches took less and less regard of natural philosophy. Confessionalized philosophy had tried to integrate philosophical insights and religious demands, but in many places lost its appeal. Physico-theology attempted a new synthesis, and to a certain extent it was very successful. However, it did not replace either the purely empirical work of the scientists or the message of the orthodox preachers. It was just a third approach next to the others.

As a result, the reactions to earthquakes became highly fragmented, as the examples in this chapter will show. To some extent, different approaches followed the demands of specific genres. Newspapers, learned journals, and sermons all fulfilled clear-cut functions. But the reactions to a specific earthquake in a specific context do not follow a fixed template. Confessional boundaries became less important. On the other hand, approaches sometimes varied considerably along national boundaries. In the following, I will discuss the reactions to several earthquakes that happened in the years around 1700. For northern (Protestant) Europe, this comes down mainly to a comparison of reactions in different countries. As to the Catholic world and Italy, apart from discussing in some detail one specific earthquake, we also have to consider some more general points.

Churches, saints, and the Italian earthquake of 1703

The traditional Catholic response to earthquakes was challenged along diverse lines. Traditionally, priests had emphasized the redemptive and saving powers of the Church. In time of peril, people were supposed to turn to the sacred rituals and objects of the Church, not just for spiritual solace, but also for mere survival. Many legends related how during an earthquake holy people remained unharmed and how church buildings remained standing amidst the general ruin. More recent reports gave similar examples. Burgos relates that when during the 1693 earthquake the vault of the prison of Palermo collapsed and the chapel was destroyed, not only the Holy Sacrament was not damaged, but even a small bird in its cage was not hurt and found back alive amidst the ruins. This caused great amazement and made people believe "that the piety practiced by the supervisor of the clergy in assisting in bringing this Host to the sick, has saved this city from complete destruction".[1]

As long as people got most information on earthquakes from ancient histories and pious legends, faith in the protective nature of the sacred was not difficult to uphold. However, with the swelling of the stream of factual information, the traditional view became increasingly difficult to uphold, for in reality, hardly any place is less safe during an earthquake than a church. Such tall structures are especially liable to come down. People who seek protection in churches are in grave danger of being killed and histories of seventeenth-century earthquakes do bear this out.

Agatio Di Somma wrote on the Calabrian earthquake of 1638 that the greatest carnage was in the churches. The quake occurred just at the time of service and churches were packed, as it happened to be the day before Palm Sunday. He specified that most victims were elderly people, women, and children, who could not flee in time.[2] Some decades later, in 1672, the Italian city of Rimini was struck by an earthquake in the Holy Week, on the early morning of Maundy Thursday, while many people were in church doing their devotion. The printed report admits that these circumstances caused the number of victims to be particularly high: many churches collapsed, killing the worshippers. Still, the author tried to uphold the providential view: the archbishop and the governor, who also attended service, were "miraculously" saved, and among the clergy the number of victims was remarkably low: only three priests, two of them of exemplary lives.[3]

Especially shocking were the events on Sicily in 1693. After a severe earthquake on January 9, most inhabitants of the city of Catania had fled into the fields. When the tremors abated, they returned. As usual in the circumstances, processions and ceremonies were organized. On January 11, a large number of people congregated in the cathedral to accompany the relics of St. Agatha. The major earthquake struck "exactly when the shrine with the relics was being opened and the people with loud voices implored the divine mercy".[4] The cathedral collapsed, killing nearly the whole congregation; a few notables saved themselves in the sacristy. Many others, who had assembled for the procession in the city's narrow streets, were killed as well when the adjacent buildings collapsed. It was obviously very hard to give a theological justification of these events. In faraway Germany, one author suggested that the disaster was a divine retribution for the abuse of offering church asylum to criminals, but only after painting the whole scene in all its horror.[5] It is doubtful that even many Protestants felt convinced and the explanation seems rather a sign of embarrassment.

To what extent such events did actually have an impact on popular feeling remains a subject of speculation, but the availability of such reports in printed form did present an alternative to the traditional legendary narratives. Among educated and informed people, including clergy, it must have given some pause. Other facts as well did not always agree with the traditional narrative. Ruffo noted that in 1726 amidst the ruins of Palermo, a long row of brothels remained standing. He explained that the fact was generally attributed to the amazing goodness of God. The prostitutes afterwards did public penance.[6]

A somewhat more pragmatic attitude seems to have crept in. In Sicily in 1693, when the earthquake started, it was proposed that the monks should leave their cloisters and convents and seek safety in the open space. They refused, however. They only consented to go into the yards, within the convent walls.[7] Thirty years later at Palermo, the situation was different. The Franciscan friar Salvatore Ruffo related how he and the other friars had initially taken refuge in the court-yard of their convent. They stayed there praying, afraid that the marmor columns of the building might collapse. The common people, however, did not leave them in peace. They insisted that the churches be opened, for prayer and confes-sion. The clergy were clearly aware that church buildings were dangerous places during an earthquake and wanted to keep them closed. However, "because of the continuous clamouring of the assembled people", as Ruffo wrote, they finally gave way. All but two of the churches in the city were opened for confession and absolution. Only the Dominicans and the Jesuits heard confession outdoors.[8]

It is very hard to claim any causal connection, but it strikes that about this same time, the Church increasingly told people to seek help from the saints rather than from sacred buildings or sacramentalia. Of course, saints had been invoked since the first centuries of Christianity, but the association of certain saints spe-cifically with earthquakes appears to be a relatively recent phenomenon. This was not an age-old popular belief, but the result of active propaganda by leading churchmen, with the purpose of comforting believers and strengthening and unifying the Church in an era of mounting challenges.

Some of these cults were purely local, such as in Spoleto, where the cult of Saint Ponziano took hold. The most important saint however became Saint Emygdius, who up to that time had not been associated with earthquakes. His cult had been mostly confined to the city of Ancona. However, when in the central Italian earthquake of 1703, Ancona remained largely unharmed, this was attributed to the intercession of its patron saint, and this considerably enhanced his reputation. As a result, the cult of Saint Emygdius as a protector against earthquakes spread far and wide, especially in southern Europe.[9] In the German lands, the role of protector against earthquakes fell to Saint Alexius, after a severe earthquake in 1670 which occurred on his feast (July 17). So, after an earthquake in 1737, the city of Rastatt put up a statue of Saint Alexius in the marketplace "to fend off the terrible earthquakes".[10]

Some other authors propagated Saint Filippo Neri, the sixteenth-century founder of the congregation of the Oratory, as a saint to be invoked during earthquakes. This had the backing of the highest ranks in the Church, to wit, Pietro Francesco Orsini, who would become Pope in 1724 under the name of Benedict XIII. (He is often regarded as the most enlightened Pope of the century.) A pamphlet published at Rome in 1688 rehearses how Orsini, at that time archbishop of Benevento, had been saved in the earthquake of 5 June of that year by the intercession of the saint. The pamphlet mainly consists of an official testimony by Orsini himself, as well as some further attestations. Orsini relates that he was in his palace when it collapsed. A heavy cabinet which contained his collection of pictures of Neri fell over him and protected him, showering him with images of the saint. He was not seriously injured and was rescued after a few hours. Orsini further relates that most of the clergy and the ecclesiastical buildings were saved. He explicitly draws a comparison with the earthquake of Antioch of 587, when only the house of the bishop remained standing.[11]

In the reactions to specific earthquakes, traditional and more recent tendencies came together. As an illustration, we will look at the earthquakes in central Italy on January 14 and February 2 1703. Apart from the serious damage in the epicentres, the quakes were also felt, though less severely, at Rome, the heart of Roman Catholicism. This was unusual and partly for this reason, the event drew widespread attention, both from philosophers and clerics. The reactions varied from physical considerations to miracle stories, as well as the usual pamphlets based on official reports of damages and victims by the authorities.

Giogio Baglivi, a physician from Rome and a prominent member of the republic of letters, wrote in Latin a short history, which he combined with some considerations he had written earlier on subterranean fire. He also added some of the official reports. Still later, he supplemented this with a second treatise on the smaller earthquakes which followed in the next two years. Baglivi is an example of the empirical turn. He is not interested in spectacular stories, but focusses on the medical effects of the earthquake and details such as the weather. He takes these partly from his own observations, partly from witnesses, and partly by going back to reports of earlier earthquakes. The religious element is very subdued. There are some pious phrases, but Baglivi's main reference is Seneca.[12]

Vincenzo Teloni, a nobleman from Viterbo, published a treatise on earthquakes in the vernacular. It follows an Aristotelian framework. Teloni mostly discusses the various standard topics but also includes a few sections on medical aspects. He appears clearly influenced by Kircher' theories and claims that the subterranean *spiritus*, which according to Aristotle causes earthquakes, probably has its origin in subterranean fires. Those fires have been put in the earth both for a natural and a moral purpose. Naturally, the subterranean fire contributes to the generation of metals, the origin of hot springs, and the like. Morally, it is the instrument used to torture the damned souls, Hell being situated at the centre of the earth. To make this point, Teloni refers to Resta's meteorology, as well as to the sermons of Saint Bernardino of Siena.[13]

Another description was written by Francesco Grimaldi, a priest, who tried to combine religious and philosophical insights. He started with a description of what had happened in the various places. This part is fairly traditional. He relates the damage and the number of victims, but is mostly interested in what happened to individual people, giving many edifying examples. While dismissing astrological or other idle predictions, he is eager to relate miracles at saints' statues or related by monks.[14] He then moves to a discussion of natural causes, following the Aristotelian theory. However, he emphasizes that nature is only a servant. Earthquakes do not happen without a special command of God. "This is also attested by the divine oracles, that ascribe an earthquake to God alone … That such scourges come forth from God as punishments for our sins, can be clearly taken from Holy Scripture." However, Grimaldi refused to call earthquakes a miracle, a view he ascribed to Cardano. "It seems to me that one cannot ascribe anything to a miracle [...] that happens naturally, when nature follows what has been pre-ordained by its supreme law-giver."[15]

On the other hand, there are treatises which focus exclusively on religion and are not interested in physical details. Bartolomeo Abbati, a lawyer associated with a cardinal's court, wrote a booklet, dedicated to the Pope, that presents itself as a meteorological treatise, but is mostly a list of prodigies and miracles. Only three pages, out of 24, are devoted to the earthquake proper, and these are fairly general. Interestingly, Abbati appears to emphasize the protection that the true Church offers believers. He relates that when the earthquake struck, the Pope was officiating in the Sixtine chapel, reciting the Litany of the Saints.

> When everything seemed to be about to fall down, he, unperturbed, raising his hands to heaven, his knees on the bare floor, gave such a great example of constancy, that nobody moved. What have the enemies of the Catholic faith, the blind and deranged heretics, to say to this?[16]

Abbati's treatise was published in Rome by Lucantonio Chracas. Chracas apparently felt that the market deserved something better, for the next year, he published under his own name another history, some 250 pages long. Physical aspects are completely lacking. The history of the earthquake itself is only three pages. Then follow nearly 150 pages of descriptions of processions, religious ceremonies, prayers, and pastoral letters of the Pope. The last part, over a hundred pages, consists of official reports of the damage in the province of Norcia.[17]

Whereas in Italy as a whole, the cult of Saint Emygdius got a boost from the earthquake, in Rome, most attention went to Filippo Neri. He was credited with another miraculous saving. A number of priests of the Oratory had been trapped in a collapsed building. However, they invoked the help of their patron saint and escaped unharmed. The miracle was not only elaborately narrated by Grimaldi, but also the subject of a separate pamphlet. The pamphlet mentions also how in the collapse of the church, the altar with the image of Filippo Neri remained standing. The story of the miracle is followed by some prayers to say during an earthquake. These are variants of the prayers from the stories of antiquity, as told by Baronio. There appeared at least one other publication of such a prayer in the same year, again at Rome.[18]

The heterogeneity of the reactions seems to reflect a deeper insecurity and not just be the result of the conventions of different genres. The religious world view was challenged not just by learned insights, but also by basic facts as they were divulged in pamphlets and histories. Different people sought solutions in different directions.

The central European earthquake of 1690

As described before, Lutheran ministers had been much preoccupied by the question of supernatural versus natural causes. At the end of the century, this was still a major issue, but by this time. some demonstrated a certain reluctance to appeal to God's direct action.[19] This becomes clear from the reactions to a severe earthquake that struck central Europe in 1690. The epicentrum lay near Villach in Carinthia (Austria), where considerable damage was reported, but it was felt in a much wider region. Actually, most of the reactions come from northern Germany. Factual information was quickly divulged. A letter in the journal *Miscellanea Curiosa* assumed that the information itself would already be well known. The author mostly speculated on the reasons why the quake had been felt stronger in some places than others.[20]

Several Lutheran ministers came to the fore, although this time only a few of them published sermons. Most reactions were treatises. These continued the traditional discourse in that they interpret the earthquake as a divine warning to abstain from sin and better our lives. People were especially impressed by the fact that in many places in Germany, the church towers were seen to swing. One treatise on the earthquake has the title: "The in the moving towers observed sign of God's punishment."[21] Schwimmer's disputation at Rudolfstadt also focussed on the towers. It is titled "small dissertation on the prodigious moving of the seats of the magnates". That church buildings and other public buildings throughout Germany had moved (Schwimmer offered a whole list), was a sign that no reasonable human, not even a beast, could ignore.[22] As particularly ominous was seen the fact that at this occasion in many cases the bells had sounded. This was perfectly natural of course, yet at the same time could be seen as a divine sign: the bells were used by God to call his people to repentance. Indeed, one minister asserted that God himself had sounded the tocsin for the Last Day.[23] An anonymous pamphlet that appears written by a Catholic also focusses on the ringing of the church bells. It has the title "Terrible warning-bell" and gives a list of places where the bells had swung or even sounded.[24]

In spite of the apocalyptic tenor, the popular treatises do not go so far as to see the earthquake as a divine miracle. They agree that it has a natural cause, but then, natural events can be used by God as well. This is a deviation from most of the Lutheran treatises on earthquakes in the first half of the century, but also from the disputation by Schwimmer at Rudolfstad, who declared the 1690 earthquake a direct act of God. So, these vernacular texts are more cautious than an academic work. But the fact that the ministers make no difficulty in seeing this earthquake as natural certainly does not mean that they are willing to abstain from the miraculous altogether.

One of the treatises was written by David Wendeler, doctor of theology. According to the title-page it had been presented to the community that was entrusted to him at Kemberg (near Wittenberg), so apparently it was originally a sermon. Wendeler stated that faith should not come from signs and wonders, but from God's Word. Still, one should not ignore the signs God uses to exhort us to penance. He rehearsed a number of examples of earlier divine warnings, most of them supernatural and from the Bible. He actually takes quite some effort to argue for the supernatural character of the Biblical earthquakes. He admitted that the earthquake of 1690 must have been natural, but even so, it is a sign of God's anger. God wants to bring us to penance by natural things as well. Here again, a number of examples follow, in particular eclipses and comets. In the last case, he remarkably quotes the first lemma of Newton's *Principia*, published only four years earlier.[25]

The minister Nikolas Höpffner wrote a treatise on the occasion of the earthquake as it was felt in Thuringia. The main part of his book is devoted to a chronological list of earlier earthquakes, with many miracle stories. Earthquakes announce disaster. We should not look upon them as the "naturalists and atheists" do, who feel that "they happen randomly and following the course of nature", but we should consider that God and nature do nothing in vain.[26] As to causes, he is a bit ambiguous. At the beginning, he explains that there are supernatural and natural earthquakes. The former are the most important. Some are worked by angels, used by God as instruments of his punishment. In other cases, God has natural causes run their course; there are various opinions on what these causes might entail. Most earthquakes happen in countries with a hollow, watery soil. In countries with a solid soil, like Thuringia, they happen only rarely, and need assistance of supernatural causes.[27] Still, given the fact that there had been much rain that year, Höpffner regarded the 1690 quake as natural.[28] This of course did not take away its character as a divine warning. He closes with an exhortation to do penance and expresses the hope that the Day of Judgement may be near.

Gottfried Erhard Fesske published "seven historical and theological treatises on earthquakes". The earthquake of 1690, which occasioned the work, is described in the preface but for the rest hardly mentioned. Fesske deals with earthquakes in a general way. He copies many facts and anecdotes from history works and periodicals. Still, he warns that the earthquake of 1690 does not mean anything good. One "should not ignore such a rare phenomenon or mistake it for a mere work of nature. We definitely have to believe that the hidden God wanted to give a special message with it".[29] In the following however, Fesske rejected outright the view that giving a natural explanation of earthquakes would be heretical. God is indeed the main cause, but he uses nature and the created world "as an assistant". Earthquakes have natural causes (although those in the Bible are supernatural).[30] That said, his treatises are mostly focussed on earthquakes as warnings and as calls for repentance. Earthquakes announce the Last Judgement and are presages of war, famine, and plague. Fesske gives several examples of earthquakes that happened because of the sins of the populace.

Still, the idea of a vengeful and angry God is partly abandoned. Everything God does is for the good and earthquakes too must have a purpose. Fesske also gives examples of God's goodness. The earthquake during the Passion destroyed many cities in Thracia and Bithynia, but in the Jewish land itself, even in the "blood-city" Jerusalem, no house was destroyed; "which without doubt so happened for the sake of some few pious believers, who were followers of the crucified Jesus and had recognized him as the true Messiah and Savior of the world".[31]

The Catholic author of the anonymous pamphlet appears still more influenced by the upcoming physico-theological ideas. He regards earthquakes in the first place as manifestations of the wisdom of God, showing the wise contrivance by which he has constituted the earth. In the second place, they should be seen as divine warnings, admonishing us to abstain from sin and calling us to repentance, as there is not much time left until the Last Judgement. In the third place, they are great and amazing wonder works. The author also inserts a letter by a Franciscan who survived the earthquake of Smyrna of 10 July 1688.[32]

The Lutheran ministers definitely continue the Lutheran meteorological tradition. They see the earthquake as a sign of impending disasters and of the Day of Judgement. However, in discussing the philosophical aspects, they have grown a bit more cautious than most of their confraters hundred years earlier. They agree in a general sense that earthquakes are divine warnings and that many are miraculous, but they are unwilling to see the 1690 earthquake itself, spectacular as it may have been, as a miracle.

The earthquake of 1692 in the low countries and England

Two years later, another earthquake happened in north-western Europe, which allows us to have a closer look at the mental climate in those Protestant countries. In the afternoon of 18 September 1692, the Low Countries were hit by an earthquake. The epicentre lay somewhere between Aachen and Liège, just south of the present Dutch border. Considerable damage occurred in the region and there were several casualties.[33] The quake was felt in a much wider area, in the Dutch Republic, the Spanish Netherlands, the adjoining parts of Germany, and south-eastern England. In this wider region, the damage was fairly limited, but as earthquakes are extremely rare here, the event still caused much consternation.

A pamphlet noted that in Amsterdam,

> the alarm was considerable, as everybody with faces that were pale, yes, pinched [*betrokken*] and stiff as a pot, came running from their houses into the streets, as everybody, because of the vehement trembling and punching

of the buildings, as well as their subsequent rocking and inclining to and fro, as if they would fall and collapse, feared and anticipated that they would be smothered underneath.[34]

Both in England and the Netherlands, reports mention that people felt dizzy because of the quake. Happening as it did at a time that philosophical and religious feelings were shifting, it offers an interesting opportunity to gauge people's sentiments.

There were no separate news sheets, but reports of the earthquake were included in some regular local journals, as the *Europische Mercurius*, published at Amsterdam, and the *Historische Kern*, published at Hamburg. The publisher of the *Historische Kern* included it also in an earthquake catalogue he published at the same time. This catalogue has a traditional content, demonstrating that earthquakes are divine instruments, but the final section on the quake of 1692 is rather factual, with detailed reports on what occurred in the English army camp in the Southern Netherlands and in the Dutch city of Middelburg.[35]

Scholars like Christiaan Huygens, Nicolaas Hartsoeker, and John Ray did take notice of the event, mostly wondering about its origin and direction, but the earthquake was not the subject of any major scientific work. Some of the news periodicals however did include philosophical speculations. The *Athenian Mercury* published an article by Ray wherein he suggested that the earthquake in Flanders and Holland "might be from vapours dispersed from that in Jamaica".[36] In Holland, the journal *Europische Mercurius* noted that the earthquake had been felt in many other places beside Holland as well, even in England. The latter was deemed remarkable, as the British Islands were separated from the continent.

> The philosophers, used to give natural causes of everything, managed to account for this by saying that the cause of this earthquake had been very deep, far below the deepest bottom of the sea that separates Great Britain from the rest of Europe. They added that for this same reason, it had not made itself feel heavier, as it had to shake an all too big part of the earth. Whereas to the contrary when these movements occur immediately under the first crust of the earth, all the buildings which it carries are shaken much more, even the soil commonly is opened at several places.[37]

The journal does not specify what "philosophers" it refers to, but the argument may have been pretty common. It is similar to what Hartsoeker would write a few years later in his *Principes de physique*. John Ray also would note that the caves where the earthquake had its origin were situated at very great depth. In order to shake such a vast extent of ground, the subterranean conducts had to pass under the sea. Otherwise, the enclosed vapour would have probably rent the earth someplace "and broken forth in the form of a flame".[38]

The many pamphlets and separate publications that did appear are hardly concerned with a description of the events and nearly exclusively with interpretation.

Such reactions are mostly known from England, in particular London, and from the Dutch Republic, that is, regions where damage was very limited. Interestingly, there are important regional differences. The English public may have reacted stronger because people were still under the impression of the reports of the earthquake of Jamaica three months earlier, that had wiped out the wealthy English colony of Port Royal. John Evelyn noticed the destruction of Jamaica in his diary on August 19. When on September 15 (Julian calendar) he wrote on the European earthquake (that is, a week after the event, as he himself had not noticed it, only "a servant or two, who were making my bed, and another in a garret"), this was rather an occasion to return to the events in Jamaica.[39] On the other hand, the London earthquake may have enhanced interest in the earlier earthquake at Jamaica. The letters by Heath on the latter were printed in London, "Licensed Sept. 9, 1692" (Julian calendar), that is, a day after the European earthquake.[40] Indeed, several authors discussed the two earthquakes together.

From England we have many printed sermons and religious treatises wherein the populace is called upon to repent their sins. Remarkably, we do not know of any printed reaction by a minister of the Dutch Reformed Church (although there is one by a dissenting minister). Indeed, a Dutch pamphlet complains that there were no public thanksgivings or other religious ceremonies after the quake. "Even some ministers made no mention of it at all in their sermons, still less in their thanksgivings, although others did pay heed."[41] The author compares this to the reactions that had been recorded from other places that had been struck by an earthquake, in particular Málaga and China, and concludes that the Catholics and the pagans show more devotion than many Reformed. He also takes offence that the pubs did not close and that the Amsterdam fair opened as usual. This again was different in England. John Evelyn reported that the fair of Southwark was closed down on September 13, apparently as a way to appease God by restricting immoral behaviour. He suggests that the occasion was "a puppet play or some such lewd pass-time" at the fair, wherein the earthquake of Jamaica was "prophanely and ludicrously represented".[42] We would like to know more about the play.

In the English sermons, the new scientific or philosophical views of the period have left hardly any trace. They follow an exclusively biblical and confessional template. The minister Walter Cross preached two sermons a week after the earthquake. His text was Revelations 11:13, on the earthquake at the resurrection of the witnesses, chosen because of "the probability or possibility" that the recent earthquakes and the one foretold in the Bible were the same. In the sermon, he prudently warned that we could not be sure that the present earthquake was one of those signal earthquakes expected, "until we see more of the Event". There are several more references to the Last Judgement. Cross claimed that earthquakes prepared the earth for the final conflagration. He also felt that it was only the Deluge which had made the earth liable to earthquakes.[43]

The first sermon was all on the witnesses, the second dealt in particular with the earthquake. Cross claimed that an earthquake is a prodigy, which he distinguished from a miracle, but not in a very clear way. Still, the earth would be immovable had not the weight of sin moved it.

> God would never have altered, or suffered to be altered, any of his Founda-
> tions to the worse, if Man had not altered his allegiance. That a most solid
> Body, 7000 Miles thick, should be shaken like a Rod, and shattered like a
> tottering Fence, is a prodigious work of God.[44]

Cross referred to a great number of natural causes that had been proposed by philosophers, but mainly to argue that these were insufficient as an explanation. They might be means, but none could be a total cause. The moving cause of earthquakes is God's anger over our sins.

Another sermon was preached at Reading by Samuel Doolittle. He preached on the text of Jesaja 29:6: "Thou shalt be visited of the Lord of hosts with thunder, and with earthquake." He used the occasion for a stern admonition. "Every Creature, from the highest Angel to the lowest worm, stands ready armed with a sting to avenge its makers quarrel. (...) In every part of the Elemental World, an angry God can find instruments of his vengeance." As for causes, Doolittle emphasized that "The sending of an earthquake is Gods act, he is the author and efficient cause of it." He did not deny that God could make use of angels and of natural (secondary) causes, but this should not distract us from the first cause. "If sometimes it be an Act of God, only as the God of Nature, concurring with second Causes: Yet many times it is an Act of God as Judge, punishing those that rebel against him."[45]

As to the particular earthquake of 1692, Doolittle asserted that "This is a warning and a presaging visit." Such visits are often "the fore-runner of some dreadful calamity that is near at hand". As to the question whether it was among the signs that presage the dissolution of the world, Doolittle warned that the book of Providence was closed for men, "yet a modest and humble conjecture is not unlawful". If God sends us a sign, we should not ignore such a warning.[46]

We do not have the date of the sermon by still another minister, John Shower.[47] Shower expanded his argument into a booklet of over two hundred pages before he had it printed. It dealt with the earthquakes of Jamaica and England as well as the earthquake of Sicily in the following year. It did contain a description of the earthquakes, but most of it was concerned with the theological interpretation.

Besides these sermons, there appeared an anonymous treatise by "a reverend Divine", *A practical discourse on the late earthquakes*. This is mainly a defence of prodigies in general. The author asserts that God sends us warnings by prophets, as well as by "strange signs and prodigies".[48] No doubt, the Deluge was foretold by some remarkable signs as well. Events like the earthquake are not just calamities in themselves, but also warnings of greater evils to come. The author rejects as atheists those people who claim that prodigies are just natural events. Only over halfway into his argument, he turns to the earthquake itself and the question what "this *Shaking of the Earth*, signifies and foretells". Remarkably, in explaining that the last earthquake does foretell another more terrible one, the author refers to the theory of subterranean winds: since the subterranean matter apparently has not been exhausted, it is still to erupt in a much more violent way. However, this is no necessity and can be averted by prayer, for "God can over-rule all Natural Causes." "For it is certain, that there is some great thing portended by an Occurrence of this extraordinary nature, as all wise and considerate persons have ever confess'd."[49]

Another treatise, by Robert Fleming, already announced in the title that earthquakes were "supernatural and premonitory signs". A modern historian called it "a curious mixture of fact and religious interpretation in terms of judgments".[50] The reverend Thomas Beverley appended a "brief consideration" of the earthquakes, of nearly forty pages, to a tract on repentance he happened to have in press.[51] Starting from Paul's letter to the Hebrews 12: 25–26, he set out to demonstrate how in this text the Kingdom of Christ was "shaded" by the shaking of heaven and earth. From there he came to talk about actual earthquakes. Considering that the earth is hanging, by the power of "the great Geometer of Heaven and Earth", as a ball in the air, Beverley felt that "any kind of Earthquake is no wonder; the wonder is, they are not every day, and not to the utmost".[52] As to the question whether the 1692 earthquake was natural or supernatural, he felt "That it is not to be ascrib'd to natural Causes any other way, than as the supreme Counsel and Being hath all Causes in his Hand, and at his Dispose." Still, he did talk about the constitution of the earth, be it in a rather vague way. He recognized the existence of "active Parts" that served as "Ammunition of Nature, or rather of the God of Nature." Earthquakes were fit as a shade of Christ's Kingdom, as demonstrated by the book of Revelation. The severe punishment of Jamaica (which Beverley compared to the destruction of Sodom) and the gentle admonition to England itself should be seen "as an argument to Repentance, and then, that they are a Prediction of a great Change on the World". Indeed, Beverly interpreted the earthquakes in a millenarian way. He expected the fall of the "great Antichristian State" as early as 1697.[53]

An anonymous *Discourse concerning divine providence, in relation to national judgments* was published in 1693. One might surmise that it was occasioned by the earthquakes, but it discusses judgements only in a general sense, without referring to a concrete occasion. The author refutes people who claim "all the great revolutions that are brought upon the World, to be the results of meer nature, which they set up as a distinct principle from *Providence*,..." He explicitly aims at mechanical philosophy: "'Tis a vain attempt to go about to explain and solve the appearances of nature by the *Mechanical Philosophy*" without taking account of the action of God.[54]

As can be seen, the Anglican reactions are purely theological. Unlike the practice that had grown during the earlier period, and unlike the near-contemporaneous German publications on the earthquake of 1690, they show little interest in philosophical or scientific aspects and only rarely refer to non-theological learned literature. English science had moved into new territory in this period, but English ministers preferred not to follow.

An English reaction somewhat outside the mainstream was written by Jacob de Rouffignac, a Huguenot minister who after the revocation of the edict of Nantes had fled to England and had learned English. He wrote a pamphlet of 47 pages on the question how similar the 1692 earthquake was to the central European earthquake of 1601, which had occurred on exactly the same date. Much of his description was concerned with the earlier quake. Unlike his Anglican colleagues, he paid much attention to natural causes and explanations, claiming that "Neither is this tracing, even of wonderful and unaccustomed Events, by their causes any ways displeasing to God...." Still, he maintained that many earthquakes were in part or wholly miraculous. They therefore had theological significance and announced great changes. The 1601 and 1692 earthquakes were of a mixed form, "that which hath a Natural cause, but extraordinarily, as it were increased by God, and in some sort enlarged beyond the Power of Nature".[55]

In the Dutch Republic, the established Church remained silent. Although the pious interpretation was certainly not absent, the emphasis in the Dutch reactions is more on mere curiosity.

Quickly, a market for souvenirs of the event emerged. An enterprising potter produced dishes with the inscription (in Dutch): "On 18 September 1692 there was an earthquake." Moreover, the majority of the Dutch reactions are commemorative poems. (No poems are known from England.) Three of them are by laypeople. One is a Mennonite woman who published under the name of her husband; one a young lawyer, Frans van Bergen; the third a physician, Dirk Liebergen. Their poems are rather pedestrian pieces of work, probably somewhat hastily written. Whether they are the result of an outpouring of personal feelings, or of prodding by a publisher, is of course hard to tell. They give very little factual information. Instead, they speak rhetorically about disasters in general, interpret them as divine warnings, and call for repentance. The religious message is prominent, but feels rather obligatory. The main function of these poems must have been to serve (and be sold) as a souvenir of the event. Since in the end, the earthquake caused little harm, the poets probably invoked God simply because they did not know what else to write.[56]

The fourth poem, "Subterranean alarm-bell", is more interesting. It was written by Georg Henrik Petri, Lutheran minister at Zaandam, a place north of Amsterdam, and is decidedly more ambitious than a mere souvenir. The printed work is thirty pages long, including the title page, a preface and a dedication to some local magistrates. The poem itself is accompanied by many annotations, which make up about half of the total text, and wherein Petri refers to many learned authors. Thus, Petri continued the tradition of learned didactic poetry. In spite of this show of erudition, the main thrust of the text is religious. Petri emphasized that this earthquake, as well as some others that occurred recently elsewhere, are a sure sign that the Day of Judgement is near. Lutherans had a strong tradition in interpreting phenomena in this way and it is probably no coincidence that the one clerical reaction in Holland was by a Lutheran, not a member of the official Dutch Reformed Church.[57]

One other Dutch pamphlet, in prose, I already referred to. It was published at Utrecht but deals with events in Amsterdam only. The anonymous author loudly complains about the overall lack of interest. As he explains, initially the fear was great, "but as soon as it was over, people just carried on as usual, as if nothing at all had happened". Moreover, he rejects the idea that natural causes contributed anything to the earthquake:

> How is it possible that we can give the least room to thoughts, as some people have done in the past and would still like to do about comets, that it should be ascribed to natural causes, as if these are just incidents that are here without much danger, dependent upon the constitution of the soil of our land and other accidental factors. As there may be found atheists who would like to interpret it in this way, and make the people still more careless.[58]

The author was not making things up. "Atheists" who flatly rejected that earthquakes were divine warnings actually existed. Although nobody did so in a treatise, the issue was addressed a few years later in a philosophical novel, "Sequel to the life of Philopater." The book, published anonymously but probably written by the proponent Johannes Duijkerius, was eventually banned because of its propagation of Spinozistic ideas; its publisher was even sent to jail. Large part of the novel is filled with philosophical conversations between three Amsterdam friends. The 1692 earthquake occurs during one of these meetings. After the event, the friends spent the rest of the afternoon discussing this kind of phenomena. As the author of the anonymous pamphlet feared, they indeed drew a comparison with comets, feeling it unlikely "that comets would appear because of the sins of the people, or in order to threaten any country with any punishments. Since these are nothing but very natural things, which are born from the coincidence of causes…" They agreed that the reason that some countries are more prone to earthquakes than others lies solely in the constitution of the various lands. They definitely rejected the view "as if such a land was being punished for the sins of its people, thoughts which they regarded as without foundation, prejudiced, and absolutely put together by ignorance and superstition".[59]

The differences between the various reactions, especially the reactions of the established Churches, are quite striking and hard to explain. German ministers mostly follow their established tradition, but pay a bit more heed to philosophical insights. English ministers are denouncing philosophy and use the earthquake to fortify their orthodox stronghold. Mainstream Dutch theology simply does not appear much to care. It is possible that English preachers saw more need to discipline their flock in the years after the Glorious Revolution of 1688. English divines were apparently pushing back against what they saw as dangerous tendencies, but philosophical radicalism was certainly not stronger in England than on the continent.

In any case, it is dangerous to see this variation as long-standing national differences. The gap between orthodoxy and science in England would soon be bridged (though certainly not obliterated) by the physico-theology of Ray and others. In the Dutch Republic, the indifference did not last. Half a century later, the earthquake of Lisbon in 1755 would elicit a large number of sermons and treatises by clergy, all following the orthodox view of sin, punishment, and the impending Last Judgement.[60] The divergence in the reactions mostly showed that there was no longer a view of the world that could pretend to be generally accepted. People were still looking for guidance on how to behave in this situation.

The lasting appeal of the belief in earthquakes as divine judgements, especially in England, may surprise some people. After all, the end of the seventeenth and the beginning of the eighteenth century, the "crisis of the European mind" as Hazard called it, saw major shifts in the view of the world. The world came to be seen as governed by natural laws; infractions from the world beyond were rejected as superstition. There were major debates on the significance of comets, on witchcraft, and on monsters. Earthquakes are just one other example of alleged prodigies (as we saw, some of the Dutch pamphlets clearly drew the parallel), so one might expect that belief in the prodigious nature of comets and earthquakes would decline together. However, that did not happen. There was never a major debate on the meaning of earthquakes as there was on the significance of comets and whereas in the eighteenth century the view that comets were warnings sent by God became pretty rare, the view that earthquakes were sent by God to punish us for our sins was still widely defended by theologians and philosophers alike. It was still dominant in the reactions to the Lisbon earthquake of 1755.

One point that made the debate on earthquakes different from the one on comets is the biblical nature of the phenomenon. Advocates of the view that comets had no special significance often emphasized that they are not as such mentioned in the Bible. This argument could not be made in the case of earthquakes, since the Bible prominently talks about them as divine instruments. The increasingly factual way of reading the Bible would even emphasize this religious significance of earthquakes. Exegetes emphasized the historical truth and reliability of the Bible, thereby incorporating the Bible into the world of physical facts. In living through an earthquake, one lived through part of the biblical story.

The main difference, however, is that whereas no single human has ever been killed by a comet, earthquakes are definitely dangerous. Even smaller earthquakes have a threatening and disturbing character. The rejection of prodigies did not imply the rejection of all divine intervention. People still saw God's hand in events that were not just unusual, but tangibly catastrophic. Disasters were still interpreted by most theologians as warnings and punishments sent by God, regardless of whether the causes were understood or not and whether the phenomenon was natural or man-made. Such disasters would include wars, fires, the explosion of powder-houses, inundations, plagues, catastrophic weather, and indeed earthquakes.

Recent authors have emphasized that the turn from religious to scientific explanations in the case of comets did not imply a complete secularization. Comets were still religiously interpreted, only the interpretation changed. No longer were they manifestations of God's anger, calling his people to repent. They were now manifestations of God's glory and demonstrations of his divine handiwork, a source not of fear but of admiration. The changing views thereby indicate a change in religious sensibility more generally. This is true, but only to a certain extent. Even though natural philosophers attempted to give earthquakes and volcanoes a place in a well-ordered economy of nature, concrete earthquakes were still seen as divine punishments. The idea of an angry and vengeful God was never far away.

Notes

1　Burgos (1693) 3.
2　Di Somma (1641) 129–131. See Batlle (1999) 76, for a case in 1428.
3　*Vera relatione* (1672) 4–6.
4　Burgos (1693) 10.
5　Alexander Polycarpus Winthern, quoted in Weber (2015) 71. See also *Sonderbahrer Bericht* (1695).
6　Ruffo (1727, German version) 12.
7　Nicolosi (1982) 25, referring to the history by Francesco Privitera (1695).
8　Ruffo (1727, Latin version) 4; (1727, German version) 4–5.
9　Poirier (2018) 160, 162–163. *Noticia* (1733) closes with a prayer to Saint Emygdius.
10　"Um Abwendung der entsetzlichen Erdbeben." Vogt (1993) 98.
11　*Relazione* (1688). See also Poirier (2018) 161.
12　Baglivi (1704), (1719).
13　Teloni (1703), par. 2–9. The purpose of the fire is discussed in par. 6.
14　Grimaldi (1703) 16.
15　Grimaldi (1703) 29–30, 31.
16　Abbati (1703) 22.
17　Chracas (1704).
18　Grimaldi (1703) 3–4. *Relatione* (1703); I could not make out whether the pages with the prayers were printed as part of this work, or were just bound with it.
19　See also Van de Wetering (1982), on a sermon published in Boston in 1693 by the Puritan Thomas Doolittle.
20　Schelhammer (1691).
21　Wendeler (1691).
22　Disputation Rudolfstadt (1691 Febr.).
23　Fesske (1691) 22–23. There exist of course earlier reports of the spontaneous ringing of church bells, e.g. *Declaration* (1619) 22.
24　*Erschröckliche Warnungs-Glocke* (1690).
25　Wendeler (1691).
26　Höpffner (1691) sect. 63.
27　Höpffner (1691) sect. 9.
28　Höpffner (1691) sect. 71.
29　Fesske (1691) 5.
30　Fesske (1691) 12–13, 8.
31　Fesske (1691) 17.
32　*Erschröckliche Warnungs-Glocke* (1690).

33 Gielen (1992) 86–87, gives the entry on the earthquake from the annals of the monastery of Rolduc and notes from the baptismal record book of Walhorn.
34 *Christelijke aanmerkingen* (1692) 9.
35 *Unglücks-Chronica* (1692). The last seven pages are on the 1692 earthquake. The work
   is conceived as kind of sequel to E.W.H. [Everhard Werner Happel], *Straff und Un-
   glücks Chronik* (Hamburg s.a.), which only dealt with inundations. The description of
   what happened in the army tallies rather closely with the diary of Constantijn Huygens jr. (1877) 124–125, who was present.
36 Schnurmann (2001) referring to *Athenian Mercury* vol. 8 no. 6, Saturday 17 September
   1692.
37 *Europische Mercurius* III–3 (July-Sept 1692) 219–220, quote on p. 220.
38 Hartsoeker (1696) 68–73; Ray (1713) 276.
39 Evelyn (1955) 113, 115.
40 Heath (1692).
41 *Christelijke aanmerkingen* (1692) 9.
42 Evelyn (1955) 115–116.
43 Cross (1692) 14, 17, 21.
44 Cross (1692) 16.
45 Doolittle (1692) 5–6, 12, 13.
46 Doolittle (1692) 20.
47 Shower (1693). On its origin as a sermon, see preface, iii. On Shower's treatise, see
   Gragg (2009) 33–34.
48 *Practical discourse* (1692) 14.
49 *Practical discourse* (1692) 22, 23.
50 Fleming (1693); Strake (1962) 64.
51 Beverley (1693) 121–160.
52 Beverley (1693) 132–133, 134.
53 Beverley (1693) 136, 147–149.
54 *Discourse concerning divine providence* (1693) 2, 9.
55 Rouffignac (1694) 2, 6–7. The pamphlet was published anonymously ("by J.D.R.
   French Minister"). The author could be identified thanks to an inscription by Rouffignac on the title page of the copy at the University of Oklahoma, whereby he presented this copy to Dr. Cappell.
56 van Bergen (1692); ten Kaate (1692); Liebergen (1692). The Dutch reactions to the
   earthquake are discussed in greater detail in Vermij (1996).
57 Petri (1692).
58 *Christelijke aanmerkingen* (1692) 9–10.
59 Duijkerius (1991) 174–175.
60 Buisman (1992) 79–107.

CONCLUSION

At the end of this study, let us return to the question with which we started. What brought about the changes in the ideas on earthquakes? And what tells us this about the development of the sciences generally in the early modern period?

One thing that became abundantly clear is that there was no autonomous or straightforward process of 'secularization' or 'disenchantment of the world' that drove developments. To the contrary, the sixteenth century saw a significant increase in the influence of religion on scientific and philosophical thinking. Social and political developments were determining factors here. In the period that the respective Churches were consolidating their position in society, scholarly theories were used to further religious and political ends. This led to a huge output of scholarly literature and at the same time deeply influenced popular ideas. The interpretation of earthquakes as divine warnings and punishments certainly had ancient roots, but within philosophy it was only imposed by the theologians of the Reformation and Counter-Reformation. In Germany, they could build on a large reservoir of popular fears, but in Italy they had to oppose and discard an alternative tradition established by the Renaissance humanists.

Political and religious demands are less obvious when it comes to explaining the changes from the second half of the seventeenth century onwards, although they did play a part. Here, a crucial factor appears to have been the continuous improvement of the means of communication and the proliferation of the sources of information. For the systematic study of rare and unpredictable phenomena like earthquakes, the enhanced information and communication was a major precondition. But obviously, it had a major impact on practically all branches of learning, a fact that maybe needs more recognition.

Better communication was already partly responsible for the changes at the time of the Reformation. The Church reformers would hardly have achieved their goals without their control of the printing press. For the changes in the second half of the seventeenth century, it was important that the means of communication did not remain in one single hand, of the state or the Church. Many groups seized upon them for their own purposes: scholars, merchants and the general public. This contributed to loosening the hold of 'confessionalized philosophy,' and opened the possibility of a new way of looking at the world, which to a large extent was free from the demands and constraints of official ideologies. I dubbed this the 'new empiricism'.

Evidently, the 'new empiricism' cannot be explained by one single cause. A deeper investigation will have to consider multiple factors, such as the establishment of philological methods for the study of texts by the Renaissance humanists, or the falling apart of the common Christian world view during the European Reformation. At this point, I just want to highlight the importance of this new empiricism for the study of nature, not just earthquakes. In the same way that the study of earthquakes and of meteors was no longer dictated by ecclesiastical policies, this was true for the study of many other aspects of the world. The history of ideas on earthquakes thereby serves as a healthy reminder that the history of science is not just the history of specific theories and discoveries. It also needs to take into account the changes in the underlying assumptions, mostly taken as self-evident and therefore rarely explicitly formulated, that dictated what questions were worth asking and what kind of answers were felt to make sense.

One final point is relevant for wider history. As we have seen, views on earthquakes in the second half of the century got fragmented in a way they had never been before, not simply determined by the demands of confession or genre. Partly, this was only natural, given that the proponents were living through a period of experimentation. But it was only possible because of the fact that ideas were no longer directed by one dominant player with clearly set goals. There were various niches, each with their own justification and their own ends. There is indeed not a clear movement towards secularization, science, physico-theology, or orthodoxy. All these things happened at the same time, but each had its own playing field. The history of the modern period is above all a history of lost unity.

APPENDIX

LIST OF RELEVANT EARTHQUAKES

This is in no way an exhaustive list of earthquakes – therefore, the reader is referred to the works by geologists. It just serves as a reference for the earthquakes mentioned in this book. This includes most of the major earthquakes in western Europe in this period (though not all of them), but also some minor ones, and a few earthquakes in other parts of the world. The events that were most commented upon, or that otherwise are most important to this study, have been printed in bold, irrespective of their magnitude or the destruction they wrought.

The catalogue also refers to sources and literature on these earthquakes. Again, the list is not exhaustive, but is limited to the sources referred to in this book. The source material on earthquakes is very extensive and a full bibliography would be a volume in itself. Under S are given indications of the most relevant primary sources that were written on the occasion of this particular earthquake. This includes histories and pamphlets, but in some cases also earthquake catalogues, general histories, or theoretical treatises, if they were written under the immediate impression of this earthquake. Under L are given references to the most important modern literature that deals with the earthquake. Full titles can be found in the bibliography.

1348 Jan 25: Villach (Austria).
L: Borst (1981).

1456 Dec 5: Naples (Italy).
L: Heitzman (2004); Figliuolo (1988); Schenk (2009).

1505 Jan 3: Bologna (Italy).
S: Beroaldo (1510).

1509 Sept 10: Istanbul (Turkey). Major disaster.
S: *Ernewerung* (1509).

1511 March 26: Swiss Alps.
S: M.H.S., *Von Erdpidem* (1511); text also published by Günther (1890) 235–237.

1528: Mainz (Germany).
S: Nausea (1531).

1531: Switzerland.
S: Paracelsus (1925) 395–401.

1538 Sept 29: Pozzuoli (Italy): the eruption of the Monte Nuovo
S: Porzio (1551); Toledo (1539).
L: Scarth (1999) 42–55.

1542, June 13: Scarperia, Tuscany (Italy).
S: *erschröckliche Newe zeytung* (1542).

1546 January 14: Jerusalem.
S: *Zeittung* (1546).

1564 June 6: Cattaro [Kotor], Dalmatia.
S: *Newe Zeyttung* (1564).

1564 July 20: Maritime Alps.
S: [*Bericht von einem Erdbidem*] (1564); *Grausame ... geschicht* (1564); *Warhafftiger Bericht* (1564); *Ware ... newe Zeytung* (1565).
L: Moroni & Stucchi (1993).

1568 July 26: Thuringia (Germany).
S: Widebram (1569).

1570–1571: Ferrara (Italy). An earthquake swarm that lasted several years and drew much attention.
S: *Ausszug unnd verteutschung* (1570); *Discours sur l'espouventable ... tremblement* (1570); Maggio (1571, 1575); Romeo (1587); Sardo (1586); Buoni (1571); Galesio (1671). See also Hottinger (1652).
L: Guidoboni (1984); Weinberg (1991); Martin (2011) 60–79; Solerti (1891) 91–100.

1571 February: Homburg, Hessen (Germany). Landslide.
S: L.M., *Wunderzeichen* (1571).

1577: Bern and Basel (Switzerland).
S: Ragor (1578).

1579 January 26: Central France.
S: *Discours espouventable* (1579).

1580 April 6, Streets of Dover. Severe earthquake, widely felt in England, France, and the Low Countries.

S: *Order of prayer* (1580); A. Fleming (1580); Twynne (1936); Harvey (1966); Tarlton, in Campbell (1940–1941); *Discours merveilleux* (1580); V.A.D.L.C., *Discours d'une … copie* (1580); *Discours des causes et effets* (1580).
L: Neilson a.o. (1984); Ockenden, in Twynne (1936); Kocher (1953) 111–114; Haynes (1979).

1581 July 21: Vienna, Austria.
S: Suevus (1587) 274v–289v; Rasch (1582).

1584 March 4: Yvorne (Swtizerland). Mountain collapse, buried one village.
S: "Narratio", manuscript, 1584; *Warhafftige unnd erbärmliche Zeittung* (1584); Aubéry (1585); Bullinger in Scheuchzer (1716) 128–132.

1590 September 5/15: Vienna, Lower Austria. Severe earthquake, felt throughout central Europe.
S: *Warhafftige und eigentliche Beschreibung* (1590); *Kurtzer bericht* (1590); Caspar (1591); Heidenreich (1591 and 1597); Maior (1591); Volmarius (1591); Fabricius (1592); Schweitzer (1593); Phyldius (1593); Konrad Kircher (1594); Meistergesang, in Streinz (1958).
L: Gutdeutsch a.o. (1987) (compilation of sources).

1598 December 16: Gera, Saxony (Germany)
S: Babst (1599).

1601 September 8/18: Unterwalden (Switzerland); widely felt throughout central Europe.
S: Osiander (1601); Padovani (1601); Moltherus (1601); Keckermann (1607); Burgower (1651); Beuther (1602); Cysat (1969) 882–887; disputation Altorf (1602 Nov.); Grobius, in Scheuchzer (1718) 85–87.
L: Ehmer (1988) 190–192.

1618 August: Piuro (Italian Alps). Mountain collapse, buried two villages. One of the most commented events of the century.
S: Gross (1618); Eckstorm (1620); La Mothe Le Vayer (1662) II, 715–718.
L: Scaramellini a.o. (1988) (compilation of sources); Presser (1957); Hauer (2009).

1619: Béarn, southern France.
S: *Declaration* (1619).

1627 July 30: Apulia (Italy).
S: *Gründlicher bericht* (1627); *Fernerer Bericht* (1627); Poardi (1627).

1638 March 17/27: Calabria (Italy). Major catastrophe.
S: *Warhaffte Relation* (1638); *Verdadera relacion* (1638); *Einkommender bericht* (1638); Recupito, in Stengel (1651) II, 244–259; Di Somma (1641); A. Kircher (1665, 1682).

1640 April 4: bend in the lower Rhine (Germany), also felt in the Netherlands.

S: Stalenus (1640); van Helmont (1648) 92–103 and (1683/1971) 131–142; van Buchell (1940) 105, 107–108.

1652, October: Sao Miguel, Azores. (Volcanic eruption.)
S: T.K., *Journael* (1653).

1655 March 19 etc.: Württenberg (Germany).
S: Nuber (1655); Wagner (1655).
L: Ehmer (1988) 196.

1657 April 24: southern Norway
S: Escholt (1663).

1667 April 6: Ragusa [Dubrovnik] (Dalmatia). Major disaster.
S: Travagini (1673); Stay (1747).
L: van Dam van Isselt (1907); Weber (2015) 65–68.

1670 July 7/17: Hall im Inntal, near Innsbruck, Tirol (Austria). Felt throughout southern Germany and northern Italy.
S: M.P.S. A.C., *Terra tremens* (1670); Dobrzensky (1671); Sturm (1698).
L: Ruggenthaler (2016).

1672 April 14: Rimini, Aemilia Romagna (Italy).
S: *Vera relatione* (1672); Mezzavacca (1692).

1674 February 17: Ambon (Dutch East Indies).
S: Rumphius (1675/1975).

1680 October 9: Málaga (Spain).
L: Udías (2009); Rueda and Fernández (2008).

1687 October 20: Lima (Peru). Major disaster.
S: Spanish report in Magnati (1688), 27–42.

1688 June 5: Benevento (southern Italy). Catastrophic earthquake.
S: Magnati (1688); Bonito (1691); *Relazione* (1688).

1690 Dec 4/Nov 24: Carinthia (Austria). Heavy earthquake, felt throughout a large part of central and eastern Europe.
S: *Erschröckliche Warnungs-Glocke* (1690); Wendeler (1691); Fesske (1691); Höpffner (1691); Schelhammer (1691); disputation Rudolfstad-Schwarzburg (1691 Febr).

1691 February 19–21: Slovenia.
S: Thalnitscher (1691).

1692 June 7/17: Port Royal, Jamaica. Major disaster.
S: Evelyn (1955) 111, 113–115; Heath (1692); Shower (1693); Phil. Trans.
L: Schnurmann (2001); Mulcahy (2008); Gragg (2009).

1692 September 8/18: Low Countries. Also felt in parts of England and Germany.

S: Const. Huygens (1877) 124–125; Chr. Huygens (1888–1950) XIX, 311; *Europische Mercurius*; *Historischer Kern*; Petri (1692); *Christelijke aanmerkingen* (1692); ten Kaate (1692); van Bergen (1692); Liebergen (1692); Cross (1692); Doolittle (1692); Shower (1693); Fleming (1693); Rouffignac (1694); Beverley (1693).
L: Alexandre et al. (2008); Vermij (1996).

1693 January 9–11: Sicily (Italy). One of the most destructive earthquakes in European history.
S: *Less-würdige Beschreibung* (ca. 1693); *Sonderbahrer nachdencklicher Bericht* (1695); Burgos (s.a.); Disputation Jena (1693 Febr.); Bottone (1721).
L: Condorelli (2013); Nicolosi, (1982); Weber (2015) 68–75.

1703 January 14, February 2: Norcia and L'Aquila (central Italy).
S: Grimaldi (1703); Llanos (1703); Abbati (1703); Chracas (1703); Teloni (1703); Baglivi (1704); *Relatione* (1703).
L: Fiorentino (1984).

1726 September 1: Palermo, Sicily (Italy).
S: *Relation* (1726); *Noticia* (1726); Ruffo (1727).
L: Vermij (2003).

1737: Karlsruhe (Germany).
S: Textor, in: Bernoulli (1752) 502–513.

1746 October 18: Lima (Peru). Major disaster.
S: Llano y Zapata (1747); *Individual* (1748); *True and particular relation* (1748); *Histoire* (1752).
L: Walker (2008).

1748 March 23: Valencia (Spain).
S: *Relaçaõ* (1748); Carrasco (1748).

1750, February 8, March 8: London (England).
S: Stillingfleet (1750); Hales (1750).
L: Kendrick (1956) 11–44; Rousseau (1969); J.G. Tayor (1975) 11–16.

1755, November 1: Atlantic Ocean, off the coast of Morocco. Commonly referred to as the earthquake of Lisbon, this is the strongest earthquake in European history and a major catastrophe. Literature on this earthquake and the reactions it elicited is very extensive and cannot be summarized here. See a.o. Kendrick (1956); Loeffler (1999); Weber (2015) 109–197. It is not dealt with in this book.

SOURCES

Archives and manuscripts

Leiden, university library, Clusius correspondence. https://clusiuscorrespondence.huygens.knaw.nl/edition.

Wolfenbüttel, Herzog August Bibliothek, 64.24 Extravag. ("Narratio de gravi terrae motu in Helvetia apud Bernates ex literis Berna datis die 27 Martij. Anno 1584.")

Printed

Anonymous (in chronological order)

Ernewerung und schier unerhört geschicht In der Stat Constantinopel und beyligenden gegenten (no place, no date [1509]). (Nürnberg, Germanisches Nationalmuseum.)

M.H.S., *Vom Erdpidem* (no place, no date [1511]).

Ein erschröckliche Newe zeytung / so geschehen ist den 12. tag Junij / jnn dem 1542. jar / jnn einem Stetlein Scharbaria genent / ... (no place, 1542).

Zeittung / von einem grossen und erschrecklichen Erdbidem / so sich den XIII. Januarij / dieses gegenwertigen xlvi. jars / im Jüdischen lande / zugetragen /... (Wittenberg 1546).

[Bericht von einem Erdbidem bei Nizza in Italien] (Nürnberg 1564).

Ein grausame und erschreckliche geschicht und zeitung / der grossen und zuvor unerhörten Erdbiebung / in des hertzogen von Sophoien Lande... (Prague 1564).

Newe Zeyttung / kurtzer Bericht / so geschehen (...) antreffende die grausame unnd ungestümme Zerstörung der Statt Cattaro / welche durch einen Erdbidem den 6. Tag des Brachmonats / in disem 1564. Jars zerstört... (Nürnberg 1564).

Warhafftiger Bericht von ellendem verderben viles volcks / so durch erschrockenliche Erdbiden / nit weit von der Statt Nizza in Liguria /... (Dillingen 1564).

Ein ware / Erschröckenliche / und Erbermmliche Newe Zeytung / von den Siben Stetten / so Gott der Allmechtig / durch ain unerhörten / Grausamen Erdbidem / alles in grundt verderbt /... (Augsburg 1565).

Discours sur l'espouvantable et merveilleux tremblement de terre advenu à Ferrare, & sur l'inondation du Pau audit lieu (Lyon 1570).

Ein Ausszug unnd verteutschung eines Brieffs vom 21. November / diss 70. Jars ... (Augsburg 1570).

L.M., Wunderzeichen Eines erschrecklichen seltzamen Erdbidems ... Geschehen diss 1571. Jars /
im Hornung / bey Homburg auff der Ohm / im Landt zu Hessen (Frankfurt am Main 1571).

Discours espouventable de l' horrible tremblement de terre advenu és villes de Tours, Orleans &
Chartres, le Lundi xxvj iour de Janvier, dernier passé, 1579 (Paris, s.a. [1579]).

Discours d'une merveilleuse et veritable copie du grand deluge & debordement de la mer ... le vi. &
vii. jour d'Avril, 1580 (Paris 1580).

Discours merveilleux et effroyable de grand tremblement de terre, advenu és villes de Rouen, Beau-
vais, Pontoise, Mantes, Poicy, Saint German en Laye, Calais & autres endroicts de ce Royaume.
Avec le traicté des Processions, & prieres publiques qui ont esté faictes le vi. iour d'Avril. 1580
(no place, 1580).

The order of prayer, and other exercises, vpon Wednesdayes and Frydayes, to auert and turn Gods
wrath from vs, threatned by the late terrible earthquake (London 1580).

V.A.D.L.C., Discours des causes et effects admirables des tremblemens de terre: contenant plusieurs
raisons & opinions des Philosophes (Paris 1580).

Warhafftige unnd erbärmliche Zeittung / von dem Erschröcklichen wunder und Fewrzeichen / Erd-
bidem / welche sich dises 84. Jahr / hin und wider begeben und zugetragen hat. Sonderlich imn
Schwitz / Berner gebiet / ein Dorff Iuorna genandt / ... (Strassburg 1584).

[Peri seismou] Orphei seu Mercurii Termaximi prognostica à terrae motibus (translated by J.Ant.
Baif) (Paris 1586).

Ein kurtzer bericht von dem sehr grossen und erschrecklichen Erdbeben / dieses jtzige 1590. Jahr den
15. Septemb. geschehen / durchs gantze Königreich Böhemb / Merhern / Osterreich / und son-
derlich in der Keyserl: Stad Wien und umbligenden örtern ... (Frankfurt an der Oder 1590).

Warhafftige und eigentliche Beschreibung / dess grewlichen / erschrecklichen und unerhörten Erdbi-
dems / so geschehen zu Wien in Oesterreich... (Cologne 1590).

Warhafftige und erschröckliche newe Zeytung / von dem gewaltigen und erschröcklichen Erdbeben /
Wassergüss / und Wunderzeichen / so über die Statt Reffel geschehen ist / den 9. tag November
inn disem 1590. Jar ... (no place ["Erstlich gedruckt zu Tantzgau"] 1590).

Commentarii collegii Conimbricensis, societatis Jesu, in quator libros de coelo, meteorologicos, &
parva naturalia, Aristotelis Stagiritae (fourth edition in Germany, Cologne 1600).

Warhafftige newe zeitung von dem Erschrecklichen Erdbidem / Fewrregen und Ungewitter zu Con-
stantinopel... (Magdeburg 1603).

Vollständige Kirchen- und Haus Music/ Darinnen ausserlesene Gesänge/ Psalmen und Hymni
(first edition Görlitz 1611, ninth edition, Breslau, s.a.).

Declaration de ce qui s'est passé sur le restablissement de la religion catholique, apostolique, & Ro-
maine, au pays de Bearn. Avec le discours du tremblement de terre, & autres prodiges qui y sont
arrivez (Lyon 1619).

Fernerer Bericht / von dem inn Welschland erschröcklichem Erdbidem... (Nürnberg 1627).

Gründlicher bericht von dem erschröcklichen Erdbidem so sich den 30. Tag Julij dieses 1627. Jahrs
begeben / da in Puglia in Welschland 6. Stadt und andere Oerter zum theil versuncken...
(Nürnberg 1627).

Kurtze Erzehlung wie in der Statt Mecha in Arabia gelegen / vor drey Monaten dieses 1630.
Jahrs / ettlich hunder Häuser (...) in einem Erdbidem und darauff folgenter Wolckenbruch unter
gangen (no place, 1630).

Verdadera relacion del espantable terremoto, sucedido a los veynte y siete de marco de 1638 (...) en la
Provincia de Calabria Citerior y Ulterior ... (Barcelona 1638).

Warhaffte Relation und gewisse Nachricht von den erschrecklichen grossen Erdbeben / so sich in
den Provincien / Calabria / Citra und Ultra den 27. Martii Anno 11638 vorgegangen ... (no
place, 1638).

Einkommender bericht und Continuation Von dem vergangnen Erschröcklichen Erdbeben in Nieder
Calabrien (...) den 27. Martii, Anno 1638. ... (Breslau, no date).

T.K., Journael, *of kort en waerachtigh verhael, van't geene onlanghs is geschiet in't eylandt St. Michiel, een van de Vlaemsche Eylanden. Vervattende de wonderbarerlijcke, vreemde, noytgehoorde ende langhdurighe aertbevingen, in den iare 1652. in de maent october* (Amsterdam 1653).

M.P.S. A.C., *Terra tremens. Die zitterende oder bebende Erde. Einfältig doch klar und deutlicher Bericht / Was Erdbeben seyen? Woher sie kommen?* ... (Nürnberg 1670).

Vera relatione del terremoto seguito nella Romagna, e Marca il Giouede Santod 14. Aprile del corrente Ano 1672. à hore 22 (Bracciano 1672).

Relaçam nova, e verdadeira, tirade da copia da carta qua hum religioso da ordem dos prégadores, que se acha em Argel em cativeiro, escreva as Illus.mo & R.mo Senhor D.F. Alfonso Henriques de S Thomas, bispo de Malega, em que lhe dá conta dos continuados tremores da terrra ... este presente anno de 1673 (s.l. 1673).

Relazione de i prodigii operati da S. Filippo Neri nella persona del Cardinal Orsini arcivescovo di Benevento, in occasione che rimase sotto le rovine delle sue stanze nel terremoto, che distrusse quella città à 5 di Giugno 1688 (Rome 1688).

Die erschröckliche Warnungs-Glocke / wodurch der langmüthig und gütige GOtt die böse Welt vor der bevorstehenden schweren Sünden-Straffe durch ein entsetzliches Erdbeben / welches dem 24. Novembr. dieses ... 1690sten Jahres in Teutschland an vielen Orten gemercket worden / väterlich gewarnet und zu wahrer Busse ermahnet... (St. Annaberg 1690).

A practical discourse on the late earthquakes: with an historical account of prodigies and their various effects (London 1692).

Christelijke aanmerkingen op de sware aardbevinge, geweest den 18. September 1692... Neffens een kort verhaal en overweginge van de 8. schrikkelyke aardbevingen, die'er nu in 35 jaren herwaarts op verscheide plaatsen, gevoelt en besuurt zijn (Utrecht 1692).

Historischer Kern oder kurtze Chronica der merckwürdigsten Geschichte des Jahres 1692 (Hamburg 1692).

Unglücks-Chronica vieler grausahmer und erschrecklicher Erdbeben... (Hamburg [1692]).

A discourse concerning divine providence, in relation to national judgments (London 1693).

Less-würdige Beschreibung der hiebevor sehr schönen/ mächtig und uralten Mittelländischen Meer-Insul oder hispanischen Königreichs Siciliae (Nuremberg, no year, ca. 1693).

Sonderbahrer nachdencklicher Bericht / von dem erschröcklichen Erdbeben / so in Sicilien vor weniger Zeit geschehen (no place, 1695).

Relatione dun' miracolo fatto dal glorioso S. Filippo Neri in preservatione di tutta la congregatione dell' Oratorio di Norcia ... nelle presenti ruine de terremoti sentiti in questo anno 1703 (Rome 1703).

J.F.M.M., *Noticia de destruiçao de Palermo, cabeça do Reino de Sicilia, causada pelo horrivel terremoto que padeceo no noite de primeiro de Setembro do anno de 1726* (Lissabon 1726).

Relation von dem entsetzlichen Erdbeben, welches den 1. Sept. an. 1726 in dem Königreich Sicilien in der Stadt Palermo geschehen, wie solches von einem guten Freund nach Neapolis beschrieben worden. Nach dem zu Wien gedruckten Original (no place no year [1726]).

Noticia do fatal terremoto succedido no reyno de Napoles em 29. de Novembro de anno de 1732. Tirada de cartas fidedignas escritas de Italia (Lissabon 1733).

Schmertzlich betrübte Nachrichten / von denen in diesem 1733. Jahr (...) grossen Wasser-Fluthen / entsetzlichen Erdbeben auch Unglücks-Fellen (no place, 1733).

A true and particular relation of the dreadful earthquake which happen'd at Lima ... on the 28th of October, 1746... , translated from the Spanish (London 1748).

Individual, e verdadeira relaçaõ da extrema ruina, que padeceo a cidade de reys Lima, capital do reyno do Peru, cum o horrivel Terremoto, acontecido em a ucrite dia 28. de Outubro de 1746., e da total assolaçaõ do presidio, e porto de Calhao... (Lissabon 1718 [=1748]).

Relaçam do lamentavel, e horroroso terremoto, que sentio, tra noute do ultimo dia de mez de Março para o primeiro de Abril de 1748 a Ilha da Madeira, extrahida de outra, que veyo de Funchal, escrita a 17 de Abril do mesmo anno (Lissabon 1748).

Relaçaõ do formidavel, elastomoso terremoto succedido no reino de Valença no dia 23 de Março deste presente anno de 1748 pelas 6. horas, e tres quartos da manhã (...) conforme as noticias communicadas até o dia 27 do mesmo mez ao Capitaõ General, Arcebispo, & Intendente , e as que successivamente vaõ chegando ã Corte de Madrid, de donde se communicaraõ a este de Lisboa (Lissabon 1748).

A chronological and historical account of the most memorable earthquakes that have happened in the world ... (Cambridge 1750).

Histoire des tremblemens de terre arrivés à Lima et autres lieux; avec la description du Perou, et des recherches sur les causes phisiques des tremblemens de terre, par M. Hales... & autres phisiciens, translated from the English (2 vol., The Hague 1752).

Periodicals

Amsterdamsche Courant, on: http://kranten.kb.nl
Acta Eruditorum
Europische Mercurius
Journal des Scavans
Miscellanea Curiosa
Philosophical Transactions

Academic disputations (in chronological order)

(N.B. Later editions in volumes with separate titles are not included here.)

Rostock 1596: Nicolaus Andreas Gran (praeses), Ulfo, son of Christophorus Andreae à Stefla, (resp.), *De meteoris, disputationes quatuor* (printed Lübeck 1596).

Marburg in Hessen 1599: Rudolph Goclenus (praeses), Johannes Leuchterus (resp.), *Disputatio duplex: altera physica, de meteoris ignitis puris; altera ethica, de fortitudine.*

Ingolstadt 1602 Sept 25: Joannes Dannemeyr (praeses), Fridericus Altstetter (resp.), *Assertiones philosophicae de impressionibus meteorologicis.*

Altorf 1602 Nov: Nicolaus Taurellus (praeses), Georgius Salmuth (resp.), *Theses de terrae motu.*

Ingolstadt 1605 March 18: Conradus Reihing (praeses), Georgius Mayer (resp.), *Assertiones philosophicae de coelo et impressionibus meteorologicis.*

Wittenberg 1606 Jan. 8: Martinus Taurellus (praeses), Jacobus Werenberg (resp.), *Disputationum meteorologicarum tertia, de ventis.*

Wittenberg 1606 [the title page says 1506], 15 Kal. Nov. [Oct. 18]: Aegidius Strauchius (praeses), Nicolaus Elerdus (resp.), *Disputatio physica de meteoris.*

Wittenberg 1607 April 4: Tobias Tandler (praeses), Johannes Stille (resp.), 'De meteoris subterraneis. De terraemotu, fontium, fluminum, & maris ortu, causis & accidentibus', in: Tobias Tandler, *Dissertationum meteorologicarum enneas* (Wittenberg 1607) number IX.

Wittenberg 1608 Febr 27: Aegidius Strauchius (praeses), Christophorus Schneiderus (resp.), *Disputatio physica de terrae motu, ex 2. Meteor. cap 7 & 8.*

Giessen 1624, Johannes Henricus Tonsor (praeses), Johannes Philippus Hugius (resp.), *Disputatio physica de terrae motu.*

Freiburg in Breisgau 1627 April 21: Georgius Reininger (praeses), Conradus Peutinger (resp.), *Doctrinae peripateticae compendium ex libris meteorologicis collectum.*

Strassburg 1642 Aug 25: Jacobus Valentinus Espichius (praeses), Georgius Klein (autor & resp.), *Dissertatio physica de terrae motu.*

Leipzig 1648 Jan 25: Christian-Friedrich Franckenstein (praeses), Johannes Mullerus (resp.), *Ad caput XXIIX. L. II. Noct. Attic. A. Gellii, exercitatio physica de terrae motu.*

Zürich 1651: Joh. Henricus Hottinger (praeses), Johannes Zollicofferus (resp.), *Quaestiones quaedam de terraemotu ex Hebraeorum atque Arabum scriptis erutae.*

Helmsted 1656: Henricus Rixner (praeses), Ernestus Andreas Stallknecht (resp.), *Disputatio physica de ignibus subterraneis.*

Wittenberg 1664: S. Kirchmajer (praeses), Michael Gundelsheimer (resp.), *Disputatio meteorologica de ostentis insolentibus aeriis.*

Jena 1672 May: Casparus Posnerus (praeses), Petrus Brand (auctor et resp.), *Dissertatio physica, de terrae motibus, & seorsim illis, qui morte & resurrectione Christi acciderunt.*

Franeker 1677: Johannes Wubbena (praeses), Vitus Ringers (autor & resp.), *Disputatio philosophica de terrae motu.*

Jena 1683 Nov 16: Jo. Andreas Schmidt (praeses), Christophorus Weisse (autor-resp.), *Terrae motum tempore passionis Christi observatum.*

Bremen 1684 May 3: Johannes Eberhard Schweling (praeses), Luderus Abraham (autor ac resp.), *Memorabilium ab orbe condito in nostra tempora usque terrae-motuum historias physicè explicatas.*

Uppsala 1686 May 17: Petrus Lagerloos (praeses), Ioannes a Weliin (resp.) *Disputatio de terrae motibus.*

Rudolfstad-Schwarzburg 1691 Febr: Joann. Michael Schwimmer (praeses), Ludovicus Fridericus Haller (resp.), *Dissertatiuncula de prodigiosè motatis magnatum sedibus, totius Germaniae, quod contigit IIX. Cal. Decemb. A.C. M.DC.XC.*

Utrecht 1692 March 29: Rodolphus Leusdenus (cand.), *Disputatio philosophica inauguralis, de terrae-motu.*

Jena 1693 Febr: Joh. Paulus Hebenstreit (praeses), Philippus Georgius Luck (auctor et resp.), *De horrendo terrae Siculae motu nuper exorto.*

Duisburg 1698 July: Fridericus Godefridus Barbeck (praeses), Theodorus Henricus Meyer (resp.), *Disputatio philosophica de igne aereo et subterraneo ejusque admirabilibus effectis pluribusque meteoris.*

Uppsala 1702 Oct. 4, Haraldus Vallerius (praeses), Petrus Gudhemius (resp.), *Disputatio physica de motibus terrae.*

Gdansk 1728 April 16: Jo. Adam Kulm (praeses), Constant. Ludovicus Wahl (resp.), *Exercitatio physica, de terrae-motibus.*

Altona 1741 Sept 13: Georg. Christ. Maternus de Cilano (praeses), Joachim Pieter (resp.), *Dissertatio physica de terrae concussionibus anno MDCCXXXVIIII in Anglia observatis.*

Utrecht 1747 Febr 22: Jacob Odé (praeses), Petrus Schagen (auctor), *Quaestio naturalis de terrae-motu.*

Other

Abbati, Bartolomeo, *Epitome metheorologica de' tremoti, con la cronologia di tutti quelli che sono occorsi in Roma...* (Rome 1703).

Achilles, Alexander, *Grund-Ursachen der Erdbebung/ oder gewaltigen Bewegungen der Erden und des Meeres/ wie auch der Erztze und Mineralien in der Erden Beschaffenheit/ und daher entspringenden warmen/ sauren und süssen Brunnen, und denn des Regens und Schnee/ auch Tau und Nebels Natur und Eigenschafften* (Berlin 1666).

Agricola, Georgius, *De natura eorum quae effluunt ex terra libri IIII* (Basel 1546).

id., *De ortu et causis subterraneorum libri V* (Basel 1546).

Ailly, Pierre d', *Tractatus super libros metheororum: de impressionibus aeris* ([Leipzig] 1506).

Albertus Magnus, 'De meteoris lib. IV', in: *Opera*, II, Petrus Jammy ed. (Lyon 1651).

id., *De meteoris* (Venice 1495).

id., *Ausgewählte Werke*, III: *Schrifte zur Geologie und Mineralogie*, I, G. Fraustadt and H. Prescher ed. (Berlin 1956).

id., *Opera omnia*, V-2, *De natura loci, de causis proprietatum elementorum, de generatione et corruptione*, Paul Hossfeld ed. (Münster 1980).

Albertus de Orlamunda [?], *Summa philosophie naturalis Alberti Magni* ([Leipzig] 1502) (a.k.a. *Philosophia pauperum*).

Ammianus Marcellinus, *Römische Geschichte* [Res gestae, Latin and German], Wolfgang Seyfarth ed., I, *Buch 14–17* (Darmstadt 1968).

Aretius, Benedictus, *Commentarii in Sacram Actuum apostolicorum historiam* (Lausanne 1579).

Aristotle, *Meteorologica*, H.D.P. Lee transl. (Cambridge (Mass.) and London 1978).

Aubéry, Claude, *De terrae motu, oratio analytica: in quo, Hybornae pagi (...) per terraemotum oppressi, historia paucis attingitur* ([Genève] 1585).

Aulus Gellius, *The Attic nights*, J.C. Rolfe transl., I (Cambridge (Mass.) and London 1984).

Auvergne, Edward d', *A relation of the most remarkable transactions of the last campaign... in the Spanish Netherlands*, Anno Dom. 1692 (London 1693).

Aventinus, Johannes (Johann Turmair), *Annalium Boiorum libri VII*, Nic. Cisnerus ed. (Basel 1580).

id., *Sämtliche Werke*, Sigmund Riezler ed. (6 vol., München 1884–1908).

Babst, Michael, *Kurtzer und warhafftiger Tractat / von dem jüngst gehörten erschrecklichen und brausenden Erdbeben...* (Freiberg 1599).

Baglivi, Giorgio, 'De terrae motu Romano, ac urbium adjacentium anno 1703', in: id., *Opera omnia medico-practica, et anatomica* (sixth edition, Lyon 1704) 501–538.

id., 'De progressione Romani terraemotus à Kalendis Martius anni MDCCIII. ad Kal. Martias anni MDCCV', in: idem, *Opera omnia medico-practica, et anatomica* (ninth edition, Antwerp 1719) 566–580.

Baronio, Cesare, *Annales ecclesiastici*, Antonius Pagius ed. (19 vol., Lucca 1738–1746).

Bartas, see Salluste.

Baudartius, Willem, *Memoryen ofte cort verhael der gedenck-weerdichste so kercklicke als werltlicke gheschiedenissen ... van den iaere 1603 tot in het iaer 1624* (second edition, 2 vol., Arnhem 1624).

Bergen, François van (Montanus), *Nederland geschud door de aardbevinge van den 8/18de van Herfstmaand des jaars 1692* (Amsterdam 1692).

Berkeley, George, 'On earthquakes', in: *The works of George Berkeley, bisshop of Cloyne*, A.A. Luce and T.E. Jessop ed., IV (London 1951) 255–256.

Beroaldo, Filippo, *Opusculum de terremotu et pestilentia, cum annotamentis Galeni* (Strasbourg 1510).

Bernoulli, Johann, *Opera omnia, tam antea sparsim edita, quam hactenus inedita, IV, Quo continentur anekdota* (Lausanne and Genève 1752).

Beuther, Johann Michael, *Compendium terraemotuum, Das ist/ kurtzer Begriff/ und gründliche Verzeichnuss/ aller unnd jeder Erdbidemen...* (Strasbourg 1602).

Beverley, Thomas, *Evangelical repentance unto salvation not to be repented of (...) upon the solemn occasion of the late dreadful earthquake in Jamaica; and the later monitory motion of the earth in London, and parts of the nation, and beyond the sea* (London 1693).

Bina, Andrea, *Ragionamento sopra la cagione de' terremoti ed in particolare de quello della terra di Gualdo di Nocera nell' Umbria seguito l'A. 1751* (Perugia 1751).

Binhard, Johann, *Newe volkommene thüringische Chronica* (Leipzig 1613).

Boaistuau, Pierre, *Histoire prodigieuses (édition de 1561). Edition critique*, Stephen Bamforth and Jean Céard ed. (Genève 2010).

Bodin, Jean, *Universae naturae theatrum. In quo rerum omnium effectrices causae, & fines contemplantur* (Frankfurt 1597).

Bonito, Marcello, *Terra tremante o vero continuatione de' terremoti dalla creatione del Mondo sino al tempo presente* (Napels 1691).

Bottari, Giovanni, 'Sopra il tremoto lezione tre', in: *Raccolta d'opuscoli scientifici e filologici*, VIII (Venice 1733) 1–102.

id., *Lezioni tre sopra il tremoto* (Rome 1748).

Bottone, Domenico, *Pyrologia topographica id est de igne dissertatio iuxta loca cum eorum descriptionibus* (Messina 1721) (First edition Naples 1692).

id., *De immani Trinacriae terraemotu. Idea historico-physica, in qua non solium telluris concussiones transactae recensentur, sed novissimae Anni 1717* (Messina 1718).

Buchell, Aernout van, *Notae quotidianae*, J.C.W. van Campen ed. (Utrecht 1940).

Buoni, Jacomo Antonio, *Del terremoto dialogo* (Modena 1571).

Burgos, Alessandro, *Siziliens Erschütterung*, transl. from the Italian (Augsburg, no year, ca. 1693).

Burgower, Johann, *Christlicher / grundtlicher Underricht von den Erdbidmen...* (Zürich 1651).

Cabeo, Niccolò, *In quatuor libros meteorologicorum Aristotelis commentaria, et quaestiones* (Rome 1646).

Camerarius, Joachim, *Norica sive de ostentis libri duo* (Wittenberg 1532).

Cardano, Girolamo, *De subtilitate libri XXI* (Basel 1611).

id., *Opera omnia* (10 vol., Lyon 1663; reprint Stuttgart – Bad Canstatt 1966).

[Carrasco, Estevan Felix], *Relacion puntual, circunstanciada de las ruinas, y extragos causados por los terremotos, que se sintieron en varias partes del Reyno de Valencia, los dias 23. de Marzo, y 2. de Abril 1748* (s.l. [1748]).

Casati, Paolo, *Dissertationes physicae de igne* (Venice 1686, used edition Frankfurt and Leipzig 1688).

Caspar, Johann, *Zwo catholische Predigen. Gehalten zu Wien in Österreich ... wider die Schröckliche Erdtbidem...* (Vienna 1591).

Castro, Christobal de, *Commentariorum in duodecim prophetas libri duodecim* (Lyon 1615).

Cesalpino, Andrea, 'Quaestionum peripateticum libri V', in: *Tractationum philosophicarum tomus unus* ([Geneva]: Eustache Vignon, 1588).

id., *De metallicis libri tres* (Nuremberg 1602).

Chambers, Ephraim, *Cyclopaedia, or, an universal dictionary of arts and sciences* (5th edition, 2 vol., London 1741).

Chracas, Lucantonio, *Racconto istorico de terremoti sentiti in Roma ... la sera de' 14 di Gennajo, e la mattina de' 2 di Febbrajo dell' anno 1703* (Rome 1703).

Cicero (Marcus Tullius Cicero), *De senectute, de amicitia, de divinatione*, William Armistead Falconer transl. (London and Cambridge (Mass.) 1971).

Cirillo, Nicola, 'Historia terraemotus Apuliam & totum fere Neapolitanum regnum, anno 1731, vexantis', in: *Philosophical Transactions* 38(428) (1733 Dec 31) 79–84.

Cross, Walter, *The summ of two sermons on the witnesses, and the earthquake that accompanies the resurrection. Occasion'd from a late earthquake, Sept. 8* (London 1692).

Cysat, Renward, *Collectanea chronica und denkwürdige Sachen pro chronica Lucernensi et Helvetiae*, I-2, Josef Schmid ed. (Lucerne 1969).

Decimator, Henricus, *Epitome meteororum, hoc est, impressionum aerearum, & mirabilium naturae operum* (Leipzig 1587).

Dell'Aquila, Matteo, *Tractatus de cometa atque terraemotu (Cod.Vat.Barb.Lat. 268)*, Bruno Figliuolo ed. (Napels 1990).

Della Porta, Giovanni Battista, *De aeris transmutationibus libri IIII. In quo opere diligenter pertractatur de ijs, quae, vel ex aere, vel in aere oriuntur. [Meteorologon] multiplices opiniones, qua illustrantur, qua refelluntur. Demum variarum causae mutationum aperiantur* (second edition, Rome 1614).

Derham, William (ed.), *Philosophical experiments and observations of the late eminent Dr. Robert Hooke (...) and other eminent virtuoso's in his time* (London 1726).

Descartes, René, *Oeuvres*, Charles Adam and Paul Tannery ed., VIII (Paris 1905).

Di Somma, Agatio, *Historico racconto de i terremoti della Calabria dall'anno 1638 fin'anno 41* (Naples 1641).

Dobrzensky de Negroponte, Joh. Jacob Wenceslaus, 'Analogia terrae motu, anno elapso in Tyroli facti, cum hypochondriacis', in: *Miscellanea Curiosa*, decuria I, annus 2 (1671) 123–124.

Doolittle, Samuel, *A sermon occasioned by the late earthquake which happen'd in London, and other places on the eighth of September, 1692* (London 1692).

Duijkerius, Johannes, *Het leven van Philopater en Vervolg van't leven van Philopater. Een spinozistische sleutelroman uit 1691/1697*, Geraldine Maréchal ed., (Amsterdam 1991).

Eckstorm, Henricus, *Historiae terrae motuum complurium, & praecipue ejus, quo Plura oppidum in Alpibus Rheticus nuper miserè obrutum & convulsum est* (Helmstad 1620).

Encyclopédie, ou dictionnaire raisonné des sciences, des arts et des métiers, XVI (Neufchatel [Paris] 1765).

Escholt, Mickel Pedersøn, *Geologia Norvegica. Or, a brief instructive remembrancer, concerning that very great and spacious earthquake, which hapned almost quite through the south of Norway: upon the 24th day of April, in the year 1657*, translated by Daniel Collins (London 1663).

Evelyn, John, *The diary of John Evelyn*, E.S. de Beer ed., V, *Kalendarium, 1690–1706* (Oxford 1955).

Fabri, Honoré, *Physica, id est, scientia rerum corporearum* (4 vol., Lyon 1669–1671).

Fabricius, Valentinus, *Chronica darinnen warhafftuge und gründliche Beschreibung / von den schrecklichen Wunderwercken und grausamen Zornzeichen Gottes den Erdbidemen...* (Erfurt 1592).

Feltmann, Gerhard, *Dissertatio de accessionibus memorabilibus immani aquarum vi vel terrae motu factis* (Amsterdam 1691).

Fesske, Gottfried Erhard, *Sieben historisch- und theologische Abhandlungen von Erdbeben (...) Nach Anleitung des ungewöhnlichen Erdbebens...* (no place [Frankfurt?] 1691).

Fleming, Abraham (ed.), *A bright burning beacon... containing a general doctrine of sundrie signs and wonders, especially earthquakes both particular and generall* (London 1580).

[Fleming, Robert], *A discourse on earthquakes; as they are supernatural and premonitory signs to a nation; with a respect to what hath occurred in this year 1692. And some special reflections thereon* (London 1693).

Fludd, Robert, *Philosophia sacra & vere Christiana seu meteorologia cosmica* (Frankfurt am Main 1626).

Francus, J., *Historicae relationis continuatio. Warhafftige Beschreibunge aller gedenckwirdigen Historien ... nechst verschienen Franckfurter Fastenmess dieses 1602. Jahres / bis auff die Herbstmess 1602 zugetragen und verlauffen haben* (Magdeburg 1602).

Frézier, Amédée François, *Relation du voyage de la mer du sud aux cotes du Chili, du Perou, et du Bresil fait pendant les années 1712, 1713 & 1714* (Amsterdam 1717).

Fritsche, Markus, *Catalogus prodigiorum atque ostentorum* (Nuremberg 1555).

id., *Meteororum, hoc est, impressionum aerearum et mirabilium naturae operum* (Nürnberg 1555).

id., *Meteororum hoc est impressionum aerearum et mirabilium naturae operum*, Johannes Hagius ed. (Wittenberg 1581).

Fromond, Libert, *Meteorologicorum libri sex* (Antwerp 1627).

idem (London 1656).

Fulke, William, *A goodly gallerye. William Fulke's book of meteors (1563)*, Theodore Hornberger ed. (Philadelphia 1979).

Galesio, August, *De terraemotu liber* (Bologna 1571).

Gassendi, Pierre, *Opera omnia*, V (Lyon 1658, repr. Stuttgart-Bad Cannstatt 1964).

Garcaeus, Johannes, *Meteorologia* (Wittenberg 1568).

idem (Wittenberg 1584).

Gemma, Cornelius, *De naturae divinis characterismis* (2 vol., Antwerp 1575).

Goclenus, Rudolph, *Disputationen zur Naturwissenschaft 1592*, Hans Günter Zekl ed., (Würzburg 2007).

Grataroli, Gulielmo, *Mundi constitutionum et tempestatum praedictiones certae ac perpetuae* (Basel 1558).

Grimaldi, Francesco Angelo, *De novo, et ingenti in universa provincia Umbriae, & Aprutij citerioris terraemotu congeminatus nuncius* (Tuderti [Todi] 1703).

Gross, Johann Georg, *Von dem erschroecklichen Undergang des Flaecken Plurs* (Basel 1618).

Guarinoni, Hippolytus, *Die Grewel der Verwüstung menschlichen Geschlechts (...)* (Ingolstadt 1616).

Hales, Stephen, *Some considerations on the causes of earthquakes* (London 1750).

Hamel, Joan Bapt. du, *De meteoris et fossilibus libri duo* (Paris 1660).

[id.], *Philosophia vetus et nova ad usum scholae accomodata* ("juxta exemplar Parisiense 1681"; 2 vol., Nuremberg 1682).

Hartmann, Philipp Jacob, 'Exercitatio de generatione mineralium, gevetabilium, & animalium in aëre, occasione anonae et telae coelitus delapsarum anno M.DC.LXXXVI in Curonia', in: *Miscellanea Curiosa*, decuria II, annus 7, appendix 1 (Nuremberg 1689).

Hartsoeker, Nicolaas, *Principes de physique* (Paris 1696).

Harvey, Gabriel, 'A pleasant and pithy familiar discourse, of the earthquake in April last', in: *The works of Edmund Spenser. A variorum edition*, X, *Spenser's prose works* (Rudolph Gottfried ed.) (third edition, London 1966) 449–462.

[Heath, Emmanuel], *A full account of the late dreadful earthquake at Port Royal in Jamaica, written in two letters from the minister of that place* (London 1692).

Heidenreich [Hedericus], Johannes, *Oratio de horribili et insolito terrae motu, qui recens Austriam vehementer concussit ... In academia Julia scripta & recitata* (Helmstedt 1591).

id., *Descriptio tristissimorum effectorum, qui horribilem terrae motum, proximis annis consecuti sunt (...) carmine elegiaco comprehensa...* (Helmstedt 1597).

Helmont, Jan Baptist van, *Ortus medicinae. Id est, initia physicae inaudita*, Franciscus Mercurius van Helmont ed. (Amsterdam 1648).

id., *Aufgang der Artzney-Kunst*, Christian Knorr von Rosenroth ed. (Sulzbach 1683; reprint Munich 1971), vol. I.

Hill, Thomas, *A contemplation of mysteries: contayning the rare effects and significations of certayne comets* (London 1571).

Hondorff, Andreas, and Wenceslaus Sturmius, *Promptuarium exemplorum. Historien und Exempelbuch...* (2 vol., Leipzig 1597 and Eisleben 1598).

Höpffner, Nicolaus, *Das erschütterte und bebende Meissen und Thüringen...* (Leipzig 1691).

Hottinger, Johann Heinrich, 'Dissertatio de terrae-motu', in: idem, *Analecta historico theologica* ([Zürich] 1652) 173–200.

Huygens, Christiaan, *Oevres complètes* (22 vol., The Hague 1888–1950).

Huygens, Constantijn, *Journaal van Constantijn Huygens, den zoon, van 21 oktober 1688 tot 2 Sept. 1696 (Handschrift van de Koninklijke Akademie van Wetenschappen te Amsterdam)* (Utrecht 1877).

Ittig, Thomas, *Lucubrationes academicae de montium incendiis* (Leipzig 1671).

Jacobus Amsfordensis [Jacobus Tymaeus de Amersfordia], *Commentaria in libros perigeneos Aristotelis. Commentarij in libros metheororum Aristotelis secundum processum Albertistarum Gymnasij quod vocant Bursam Laurentij in Colonia* (Cologne 1513).

Kaate, N. ten, *Ter errinering van de verschrikkende aard-beeving; in de Neederlanden, en elders, voorgevallen op den 18den van herfstmaand, in't jaar 1692* (Amsterdam 1692).

Keckermann, Bartholomaeus, *Contemplatio gemina, prior, ex generali physica de loco; altera, ex speciali, de terrae-motu, potissimum illo stupendo, qui fuit anno 1601 mense Septembri* (second edition, Hanau 1607).

Kircher, Athanasius, *Diatribe de prodigiosis crucibus, quae tam supra vestes hominum, quam res alias, non pridem post ultimum incendium Vesuvij Montis Neapoli comparuerunt* (Rome 1661).

id., *Mundus subterraneus in XII libros digestus* (Amsterdam 1665).

id., *D'onder-aardse weereld in haar goddelijk maaksel* (Amsterdam 1682).

Kircher, Konrad, *Zehen Predigen / von erschröcklichen Erdbidemen (...) gehalten zu Sonnberg in Nider Oesterrreich* (Laugingen 1594).

Köber, Joh. Friedrich, *Dissertatiuncula de motu terrae et eclipsi Solis circa passionem Christi, numquid de iis gentilium scriptoribus innotuerit?* (Gera 1683).

Kochanski, Adam Adamowich, 'Considerationes & observationes physico-mathematicae circa diurnam telluris vertiginem, a multis absque certis demonstrationibus assertam', in: *Acta eruditorum* vol. 4 (1685) 317–327.

Konrad von Megenberg, *Das Buch der Natur. Die erste Naturgeschichte in deutscher Sprache*, Franz Pfeiffer ed. (Stuttgart 1861).

idem, *Naturbuch / vonn Nutz / Eigenschafft / Wunderwirckung und Gebrauch aller Geschöpff / Element unnd Creaturn...* (Frankfurt am Main 1536).

Krüger, Johann Gottlob, *Geschichte der Erde in den alleraeltesten Zeiten* (Halle 1746).

[Krüger, Johann Gottlob], *Histoire des anciennes revolutions du globe terrestre. Avec une relation chronologique et historique des tremblemens de terre, arrivés sur notre globe depuis le commencement de l'ère chrétienne jusqu'à présent* (Amsterdam 1752).

La Mothe Le Vayer, François de, *Oeuvres*, (third edition, 2 vol., Paris 1662).

La Primaudaye, Pierre de, *Academie francoise* (3 vol., Paris 1593–1594).

Lefèvre d'Estaples, Jacques, *Meteorologia Aristotelis eleganti paraphrasi explanata* (no place [1512]).

Léméry, Nicolas, 'Explication physique & chymique des feux souterrains, des tremblemens de terre, des ouragans, des eclairs & du tonerre', in: *Mémoires de mathematique et de physique, tirez des registres de l'Académie Royale des Sciences.* De l' année 1700, 101–110.

Leuchterus, Henricus, *Discurss von etlichen Zeichen / welche sich biss dahero (...) am Himmel und hierniden auff Erden zugetragen haben...* (Darmstadt 1613).

Liebergen, Dirk, *Op de aard-beving, voorgevallen op den achtienden van herfstmaand, des jaars MDCXCII* ('s Gravenhage 1692).

Lister, Martin, *De fontibus medicatis Angliae, exercitatio nova & prior* (Frankfurt and Leipzig 1684).

Llano y Zapata, Joseph Eusebio de, *Carta ó diario ... en que con la mayor verdad, y critica mas segura le dà cuenta de todo lo acaecido en esta capital de el Peru, desde el Viernes 28. de Octubre de 1746. quando experimentò su mayor ruìna, con el grande movimento de tierra, que padeciò à las diez, y media de la noche del mencionado dia, hasta 16. Febrero de 1747...* (Madrid 1748).

Uria de Llanos, Alfonso, *Relazione, o vero itineraria ... per riconoscere li danni causati dalli passati terremoti sequiti li 14. Gennaro, e 2. Febraro M.DCC.III. nella provincia dell'Aquila* (Rome 1603 [=1703]).

Lonzer [Lonicerus], Joannes, *De meteoris, compendium, ex Aristotele, Plinio, & Pontano* (Franfurt 1548).

Luther, Martin, *Werke. Kritische Gesamtausgabe*, XXXXII (127 vol., Weimar 1883–2009).

Lydiat, Thomas, *Praelectio astronomica de natura coeli & conditionibus elementorum (...)* (London 1605).

Maggio, Lucio, *Del terremoto dialogo* (Bologna 1571).

id., *Discours du tremblement de terre en forme de devis*, Nicolas Le Livre transl. (Paris 1575).

Magnati, Vincenzo, *Notitie istoriche de'terremoti succeduti de'seculi trascorsi, e nel presente* (Naples 1688).

Maiole, Simone, *Dies caniculares, hoc est, colloquia physica nova & admiranda* (third edition, Mainz 1654).

Maior, Johannes, *Elegia sive votum, pro illustrissima principe Sophia (...) Item, elegia de terrae motu* (Wittenberg 1591).

Mazzotta, Benedetto, *De triplici philosophia: naturalis, astrologica, et minerali (...)* (Bologna 1653).

Melanchthon, Philippus, *Initiae doctrinae physicae* (Frankfurt 1550).

Meurer, Wolfgang, *Meteorologia*, Christophorus Meurer ed. (Leipzig 1587).

id., *Meteorologia*, Christophorus Meurer ed. (Leipzig 1592).

id., *Meteorologia*, Christophorus Meurer ed. (Leipzig 1606).

Mezzavacca, Flaminio, 'De terraemotu libellus, in quo curiosa aperitur terraemotus doctrina, & agitur de terraemotu Anni 1672', in: Gaudentius Robertus ed., *Miscellanea Italica physico-mathematica* (Bologna 1692).

Moltherus, Johannes, *Eine christliche Predigt von dem Erdbeben* (Ursel 1601).

Montanus, Arnoldus, *Gedenckwaardige gesantschappen der Oost-Indische Maetschappy in 't Vereenigde Nederland aan de Kaisaren van Japan ...* (Amsterdam 1669).

Namslerus, David, *Ausführlicher Bericht von Wassern und Wasserflutten* (Liegnitz 1608).

Nardi, Giovanni, *De igne subterraneo physica prolusio* (Florence 1641).

Nauclerus, Johannes, *Chronica ab initio mundi usque ad annum Christi nati M.CCCCC* (Cologne 1614).

Nausea, Fridericus (Friedrich Grau), *Ad sacratissium Caesarem Ferdinandum (...) super huius anni post Christum natum M.D.XXXI. & quolibet alio Cometa explorrtio* [sic] (Mainz 1531). (Contains also: *De praecipuo huius anni post Christum natum M.D.XXVIII. apud Moguntam terrae motu responsum.*)

id., *Libri mirabilium septem* (Cologne 1532).

id., *De locustis* (1544).

Nieuwentijt, Bernard, *Het regt gebruik der werelt beschouwingen* (Amsterdam 1715).

Nipho, Agostino (Suessanus), *Subtilissima commentaria in libros meteorologicorum, & in librum de misti, sive quartum meteororum* (Venice 1560).

Nuber, Georg, *Eine christliche Erinnerung von de schröcklichen Erdbidemen / welche sich dises instehende Jahr sehr vilfältig und schrecklich erzeigt* (Stuttgart 1655).

id., *Conciones meteoricae: Das ist christliche in Gottes Wort gegründte Predigten von den Zeichen / so in der Lufft und obern Elementen sich begeben...* (Ulm 1661).

Osiander, Andreas, *Kurtze und einfältige Predigt / vom Erdbidem* (Tübingen 1601).

Ozanam, Jacques, *Récréations mathématiques et physiques*, J.E. Montucla ed. (Paris 1790).

Padovani, Fabrizio, *Tractatus duo. Alter de ventis. Alter perbrevis de terraemotu* (Bologna 1601).

Paracelsus (Bombastus Theophrastus von Hohenheim), *Sämtliche Werke*, 1. Abteilung, Karl Sudhoff ed., IX (Munich 1925) and XIII (Munich and Berlin 1931).

Petrarca, Francesco, *De remediis utriusque fortunae libri II* (Rotterdam 1649).

Petri, Georg Henrik, *Onder-aardze storm-klok, door de almachtige en alleen wonder-doende hand Gods, in een onverwachte en verschrikkelyke aardbeeving ... den 18. September ... 1692* (Amsterdam 1692).

Peucer, Caspar, *Commentarius de praecipuis divinationum generibus* (Wittenberg 1553).

id., *Commentarius de praecipuis generibus divinationum* (Wittenberg 1580); idem (Servestae [=Zerbst] 1591).

Phyldius, Joannes, *Ein christliche Busspredigt bey dem schrecklichen Erdbeben zu Wien in Oesterreich ...* (Ursel 1593).

Piccolomini, Francesco, *Librorum ad scientiam de natura attinentium: partes quinque* (Frankfurt 1597).

Plinius (Gaius Plinius Secundus), *Liber II. de mundi historia*, cum commentariis Iacobi Milichii (Frankfurt 1543); (idem Frankfurt 1563).

id., *Natural history, I, Praefatio, libri I, II*, H. Rackham transl. (Cambridge and London 1979).

Poardi, Joannes V. de, *Relatio historica, terrae motus horribilis... / Von dem grossen schreckichen Erdbidem* (Augsburg 1627).

id., *Relation von dem grossen schröcklichen Erdbidem... jetzt verteutscht* (Augsburg 1627).

Pontano, Giovanni Gioviano, *Meteororum liber* (Wittenberg 1524).

id., *Liber de meteoris cum interpretatione Viti Amerbachi* (Strasbourg 1545).

Porzio, Simone, *De conflagratione agri Puteolani* (Florence 1551).

Ragor, Johann Huldrich, *Von den Erdbidem. Ein grundlicher Bericht / was dieselbige seyen / auss was ursach sie entstanden / und auff was end hin sie beschehen* (Basel 1578).

Rasch, Johann (ed.), *Von Erdbidem etliche Tractät / alte und newe / hocherleuchter und bewährter Scribenten ...* (Munich 1582).

Ray, John, *Three physico-theological discourses, concerning I. The primitive chaos and creation of the world. II. The general deluge, its causes and effects. III. The dissolution of the world, and future conflagration* (third edition, London 1713).

Reinzer, Franz, *Meteorologia philosophico-politica* (Augsburg 1709).

Resta, Francesco, *Meteorologia de igneis aereis aqueisque corporibus* (Rome 1643).

Richter, G. Frid., *De natalibus fulminum tractatus physicus* (Leipzig 1725).

Romeo, Annibale, *Dialogo, diviso in due giornate nella prima delle quali si tratta delle cause universali del terremoto ... Nella seconda ... s' assegnano cause particulari...* (Ferrara 1587).

Rottmann, Ioannes, *Meteorologiae synopsis* (Frankfurt am Main 1619).

[Rouffignac, Jacob de], *The earth twice shaken wonderfully: or, an analogical discourse of earthquakes, its natural causes, kinds, and manifold effects; occasioned by the last of these, which happened on the eighth day of September 1692. At two of the clock in the afternoon* (London 1694).

Rudolphus, Valentinus, *Zeitbüchlein darinnen gründlich ... angezogen / was ... an Kriegen / Theurenzeiten / Zeichen an Hmmel unnd Erden ... ergangen* (Jena 1580).

id., *Ausführliche nachricht von dem erschröcklichen Erdbeben, welches sich zu Palermo in Sizilien, den 1. Sept. 1726, ereignet* (Leipzig 1727).

Ruffo, Salvatore, *De horrendo terrae motu qui contingit Panormi nocte post Kalend. sept. MDCCXXVI tractatus historicus* (Leipzig 1727) (translation by G.F. Richter).

[Rumphius, G.E.], *Waerachtigh verhael, van de schrickelijcke aerdbevinge, nu onlanghs eenigen tyd herwaerts, ende voornaementlijck op den 17. February des jaers 1674. voorgevallen, in, en ontrent de eylanden van Amboina* (z.pl. 1675; facsimile reprint 1975, with a transcription and English translation by W. Buijze).

Sack, Siegfried, *Erklerung uber die trostreiche historiam des Leidens und Sterbens unsers HErrn und Heilandes Jhesu Christi...* (Magdeburg 1591).

Salluste du Bartas, Guillaume de, 'La sepmaine', in: *Works*, U.T. Holmes et al. ed., II (Chapel Hill 1938, reprint Genève 1977).

Sardo, Alessandro, *Discorsi* (Venice 1586).

Schelhammer, Günther Christoph, 'De nupero terrae motu', in: *Miscellanea Curiosa*, decuriae II, annus 9 (1690) (Nuremberg 1691) 246–249 (Observatio 144).

Scheuchzer, Johann Jakob, *Physica, oder Natur-Wissenschaft* (2 vol., Zürich 1703).

id., *Beschreibung der Natur-Geschichten des Schweizerlands* (3 vol., Zürich 1706–1708).

id., *Helvetiae stoicheiographia. Orographia. Et oreographia. Oder Beschreibung der Elementen / Grenzen und Bergen des Schweizerlands* (Zürich 1716).

id., *Hydrographia Helvetica. Beschreibung der Seen/ Flüssen/ Brünnen/ warmen und kalten Bädern/ und anderen Mineral-Wassern des Schweitzerlands* (Zürich 1717).

id., *Meteorologia et oryctographia Helvetica. Oder Beschreibung der Lufft-Geschichten / Steinen / Metallen / und anderen Mineralien des Schweitzerlands / absonderlich auch der Überbleibselen der Sündfluth* (Zürich 1718).

id., *Natur-Geschichte des Schweitzerlandes, samt seinen Reisen über die Schweitzerischen Gebürge* (Joh. Georg Sulzer ed.) (2 vol., Zürich 1746).

Schweitzer, David, *Ein christliche Busspredigt / auch gründtliche unnd ausführliche Erklärung / der erschröcklichen / grausamen und schädlichen Erdbeben ... gehalten zu Schöngraben in Nider Oesterreich ...* (Frankfurt am Main 1593).

Secinara, Filippo de, *Trattato universale di tutti li terremoti occorsi, e noti nel mondo, con li casi infausti, ed'infelici pressagiti da tali terremoti* (Aquila 1752).

Seneca (Lucius Annius Seneca), *Naturales quaestones*, Thomas H. Corcoran transl. (Cambridge (Mass.) and London 1972).

Shower, John, *Practical reflections on the late earthquakes in Jamaica, England, Sicily, Malta, &c. Anno 1692. With a particular, historical account of those, and divers other earthquakes* (London 1693).

Stalenus, Johannes, *Oratio in recentem terrae-motum* (Cologne 1640).

Stanhuf, Michael, *Sylvula complectens praecipua meteororum genera* (Wittenberg 1554).

id., *De meteoris libri duo* (Wittenberg 1578).

Stay, Benedictus, *Philosophiae versibus traditae libri sex* (second edition, Rome 1747).

Stengel, Georg, *De monstris et monstrosis, quam mirabilis, bonus, et iustus, in mundo administrando, sit Deus, monstrantibus* (Ingolstadt 1647).

id., *Opus de iudiciis divinis, quae Deus in hoc mundo exercet* (4 vol., Ingolstad 1651).

Stillingfleet, Benjamin, *Some thoughts occasioned by the late earthquakes. A poem* (London 1750).

Strype, John, *The history of the life and acts of the most reverend father in God, Edmund Grindal, the first bishop of London, and the second archbishop of York and Canterbury succesively, in the reign of Q. Elizabeth* (London 1710).

id., *Annals of the Reformation and establishment of religion, and other various occurences in the Church of England* (2 vol., London 1725).

Sturm, Johann Christoph, *Philosophia eclectica, h.e. exercitationes academicae, quibus philosophandi methodus selectior ... fideliter ac dilucidè explicatur*, I (Altorf 1698).

Suevus [Schwabe], Siegmund, *Spiegel des menschlichen Lebens* (Leipzig 1587).

Tandler, Tobias, *Dissertationum meteorologicarum enneas* (Wittenberg 1607).

Telesio, Bernardino, *De his, quae in aëre fiunt; & de terrae-motibus, liber unicus* (Naples 1570).

Teloni, Vincenzo, *De terremoti, loro cagioni, effetti, e malori, che producono, e loro cura preservativa in generale* (Viterbo 1703).

Thalnitscher, Johann Georg, 'De terrae motu Labaci, Carnioliae, die XIX Febr. Anni 1691. & subsequentibus duobus diebus', in: *Miscellanea Curiosa*, decuriae II, annus 9 (1690) (Nürnberg 1691) 423–427 (Observatio 226).

Thoum, Louys du, *Le-tremble-terre ou sont contenu ses causes, signes, effets, et remedes* (Bordeaux 1616).

Tirinus, Jacobus, *Commentarius in vetus et novum testamentum* (3 vol., Antwerp 1632).

Toledo, Piero Giacomo de, *Ragionamento, del terremoto, del nuovo monte, del aprimento di terra in Pozuolo, nel anno 1538., e dela significatione d'essi* (Napels 1539).

Torreblanca, Francisco, *Daemonologia sive de magia naturali, daemoniaca, licita, & illicita, deque aperta & occulta, interventione & invocatione daemonis* (Mainz 1623).

Torres Villaroel, Diego de, *Tratado de los tremblores, y otros movimientos de la tierra, llamados vulgarmente terremotos* (Madrid 1748).

Travagini, Francesco, *Super observationibus a se factis tempore ultimorum Taerremotuum [sic], ac potissimum Ragusiani, physica descriptio. Seu gyri terrae diurni indicium* (no place, "juxta exemplar Venetiis impressum, anno 1673").

Trithemius, Joannes, *Annalium Hirsaugiensinum* (2 vol., Sankt Gallen 1690).

Twynne, Thomas, *Thomas Twynne's discourse on the earthquake of 1580*, R.E. Ockenden ed. (Oxford 1936).

Veltkirch [Velcurio], Joannes, *Epitomae physicae libri quatuor* (Erfurt 1538).

id., *Commentarii in universam physicam Aristotelis libri quatuor* (Tübingen 1539).

Vicomercato, Francesco, *In quatuor libros Aristotelis meteorologicorum commentarii* (Venice 1565) [earlier published Paris 1556].

Vogler, Valentin Heinrich, *Physiologia historiae passionis Jesu Christi* (Helmstedt 1673).

Volmarius, Marcus, *Newe Zeittung vom schröcklichen Erdbidm ... zu Wien in Oesterreich geschehen ... sampt einer Erleuterung* (no place, 1591).

Wagner, Tobias, *Zwo ernsthaffte / scharffe Buss-Predigten / uber den fortsetzenden Erdbidem im Hertzogthumb Württenberg / und dero Orten...* (second edition, Tübingen 1655).

Wassenberch, Johann, *Die Chronik des Johann Wassenberch. Aufzeichnungen eines Duisburger Geistlichen über lokale und weltweite Ereignisse vor 500 Jahren*, Arend Mihm ed. (Duisburg 1981).

Wendeler, David, *Das an denen bewegten Thürmen erblickte Straff-Zeichen GOttes / aus der Apostel Geschichte am XVI. Cap. v. 26 ...* (Leipzig 1691).

Wendelen, Govaert (Godfried Vendelinus) (ed.), *De causis naturalibus pluviae purpureae Bruxellensis clarorum virorum indicia* (Brussels 1647).

Wellendorfer, Virgilius (from Salzburg), *Decalogium de metheorologicis impressionibus, et mirabilibus nature operibus* (s.l., s.a. [Leipzig? 1507]).

Wettersteint de Hodenstein, Leopold (pseud.), *A full account of the great and terrible earthquake in Germany, Hungary and Turkey ... as also a learned discourse of the nature, causes and kinds of earthquakes ...*, translated by Rich. Alcock (London 1673).

Wick, Johann Jakob, *Wickiana. Johann Jakob Wicks Nachrichtensammlung aus dem 16. Jahrhundert*, Matthias Senn ed. (Küsnacht-Zürich 1975).

Widebram, Friedrich, *Terraemotus, parelia, paraselenae, fasces, sagittae & faces Jenae in aëre conspectae* (no place, 1569).

Willich, Jodocus, *Isagoge in Aristotelis, Alberti Magni, et Pontiani meteora* (Frankfurt an der Oder 1549).

Wildenbergius [Cingularius?], Hieronymus, *Totius philosophiae humaniae in tres partes, nempe in rationalem, naturalem & moralem, digesto* (corrected edition, Basel 1585).

Woodward, John, *An essay towards a natural history of the earth and terrestrial bodies, especially minerals (...) with an account of the universal Deluge* (second edition, London 1702).

Ziegler, Jakob, *Grundlicher Bericht / von den natürlichen Ursachen der Erdbidmen. Samt angehenkter historischer Erzehlung / was mehrenteils darauf in unserem geliebten Vatterland erfolget* (Zürich 1674).

Zuccolo, Vitale, *Dialogo delle cose meteorologiche, in cui si dichiarano tutte le cose marauigliose, che si generano nell'aere, & alcune mirabili proprietà de fonti, fiumi, e mari, secondo la dottrina d'Aristotele con le opinioni di altri illustri scrittori* (Venice 1590).

Secondary literature

Adams, Frank Dawson, *The birth and development of the geological sciences* (Baltimore 1938).

Akasoy, Anna A., 'Islamic attitudes to disasters in the middle ages: a comparison of earthquakes and plagues', in: *The Medieval History Journal* 10 (2007) 387–410.

id., 'Interpreting earthquakes in medieval Islamic texts', in: Christoph Mauch and Christian Pfister ed., *Natural disasters, cultural responses. Case studies toward a global environmental history* (Plymouth etc. 2009) 183–196.

Alexandre, Pierre, David Kusman, et al., 'The 18 September 1692 earthquake in the Belgian Ardenne and its aftershocks', in: J. Fréchet, M. Meghraoui and M. Stucchi ed., *Historical seismology* (Heidelberg 2008) 209–230.

Almagia, Roberto, 'Le dottrine geofisiche di Bernardino Telesio. Primo contributo ad una storia della geografia scientifica nel cinquecento (1908)', in: idem, *Scritti geographici (1905–1957)* (Rome 1961) 151–178.

Alvarez, Walter, and Henrique Leitao, 'The neglected early history of geology. The Copernican revolution as a major advance in understanding the earth', in: *Geology* 38 (2010) 231–234.

Ashworth, William B., 'Natural history and the emblematic world view', in: David C. Lindberg and Robert S. Westman ed., *Reappraisals of the scientific revolution* (Cambridge 1990) 303–332.

Baratta, Mario, 'Ricerche storiche sugli apparecchi sismici', in: *Annali dell'Ufficio Centrale Meteorologico e Geodinamico Italiano*, series 2, 17 (1895) 1–37.

Barnes, Robin B., *Prophecy and gnosis. Apocalypticism in the wake of the Lutheran Reformation* (Stanford 1988).

id., *Astrology and reformation* (Oxford 2016).

Barnett, Lydia, *After the flood. Imagining the global environment in early modern Europe* (Baltimore 2019).

Battle, Carmen, 'Destructions naturelles des villes de la Couronne d'Aragon au bas Moyen Âge', in: Martin Körner ed., *Destruction and reconstruction of towns. Destruction by earthquakes, fire and water*, I (Bern, Stuttgart, Wien 1999) 67–86.

Behringer, Wolfgang, 'Communications revolutions: a historiographical concept', in: *German History* 24 (2006) 333–374.

Begemann, Christian, *Furcht und Angst im Prozess der Aufklärung. Zu Literatur und Bewusstseinsgeschichte des 18. Jahrhunderts* (Frankfurt am Main 1987).

Berlioz, Jaques, *Catastrophes naturelles et calamités au Moyen Age* (Turnhout 1998).

Blair, Ann, *The theatre of nature. Jean Bodin and Renaisance Science* (Princeton 1997).

id., 'Un poème inconnu de Jean Bodin', in: *Bibliothèque d'Humanisme et Renaissance* 54 (1992) 175–181.

Blair, Ann, and Kaspar von Greyerz (ed.), *Physico-theology. Religion and science in Europe, 1650–1750* (Baltimore 2020).

Bogucka, Maria, 'The destruction of towns by natural disaster, as reported in early modern newspapers', in: Martin Körner ed., *Destruction and reconstruction of towns. Destruction by earthquakes, fire and water*, I (Bern, Stuttgart, Wien 1999) 309–319.

Borrelli, Arianna, 'The weatherglass and its observers in the early seventeenth century', in: Claus Zittel, Gisela Engel, et al. ed., *Philosophies of technology. Francis Bacon and his contemporaries* (Leiden and Boston 2008) I, 67–130.

Borst, Arno, 'Das Erdbeben von 1348. Ein historischer Beitrag zur Katastrophenforschung', in: *Historische Zeitschrift* 233 (1981) 529–569.

Brefeld, J., *A guidebook for the Jerusalem pilgrimage in the late Middle Ages* (Hilversum 1994).

Brosseder, Claudia, *Im Bann der Sterne. Caspar Peucer, Philipp Melanchthon und andere Wittenberger Astrologen* (Berlin 2004).

Brunner, Horst, Burghart Wachinger, et al. (ed.), *Repertorium der Sangsprüche und Meisterlieder des 12. bis 18. Jahrhunderts* (16 vol., Tübingen 1986–2009).

Buisman, Jan Willem, *Tussen vroomheid en Verlichting. Een cultuurhistorisch en -sociologisch onderzoek naar enkele aspecten van de Verlichting in Nederland* (2 vol., Zwolle 1992).

Campbell, Lily B., 'Richard Tarlton and the earthquake of 1580', in: *Huntington Library Quarterly* 4 (1940–1941) 293–301.

Capel, Horacio, *Organicismo, fuego interior y terremotos en la ciencia Española del siglo XVIII* (Barcelona 1980).

Caplan, Jay, *Postal culture in Europe 1500–1800* (Oxford 2016).

Clark, Stuart, *Thinking with demons. The idea of witchcraft in early modern Europe* (Oxford 1997).

Céard, Jean, *La nature et les prodiges. L'insolite au XVIe siècle, en France* (Génève 1977).

id., 'La notion de prodige selon Cornelius Gemma', in: Hiro Hirai ed., *Cornelius Gemma. Cosmology, medicine and natural philosophy in Renaissance Louvain* (Pisa and Rome 2008) 67–76.

id., 'Les *Météorologiques* d'Aristote à la Renaissance: la paraphrase de Jacques Lefèvre d'Étaples et le commentaire de Cochlaeus', in: T. Belleguic and A. Vasak ed., *Ordre et désordre du monde. Enquête sur les météores, de la Renaissance à l'âge moderne* (Paris 2013) 29–50.

Condorelli, Stefano, 'Le tremblement de terre de Sicile de 1693 et l'Europe: diffusion des nouvelles et retentissement', in: *Dimensioni e Problemi della Ricerca Storica* 2 (2013) 139–166.

Cook, Harold J., *Matters of exchange. Commerce, medicine, and science in the Dutch Golden Age* (New Haven and London 2007).

Croke, 'Two early Byzantine earthquakes and their liturgical commemoration', in: *Byzantion* 51 (1981) 122–147.

Dagron, Gilbert, 'Quand la terre tremble…', in: *Travaux et mémoires 8. Hommage à M. Paul Lemerle* (Paris 1981) 87–103.

Dal Prete, Ivano, '"Being the world eternal…" The age of the earth in Renaissance Italy', in: *Isis* 105 (2014) 292–317.

Dannenfeldt, Karl H., 'Wittenberg botanists during the sixteenth century', in: L.P. Buck and J.W. Zophy ed., *The social history of the reformation* (Columbus 1972) 237–238.

Dam van Isselt, W.E. van, 'Eenige lotgevallen van Jacob van Dam, consul te Smirna van 1668–1688', in: *Bijdragen voor Vaderlandsche Geschiedenis en Oudheidkunde*, series 4, 6 (1907) 78–136.

Daston, Lorraine, and Katharine Park, *Wonders and the order of nature 1150–1750* (New York 1998).

Delumeau, Jean, *La peur en occident. Une cité assiégiée (XIVe–XVIIIe siècles)* (Paris 1978).

id., *Le péché et la peur. La culpibilisation en occident (XIIIe-XVIII siècles)* (Paris 1983).

Delumeau, Jean, and Yves Lequin (ed.), *Les malheurs des temps. Histoire des fléaux et des calamités en France* (Paris 1987).

Dewey, James, and Perry Byerly, 'The early history of seismometry (to 1900)', in: *Bulletin of the Seismological Society of America* 59(1) (February 1969) 183–227.

Draelandts, Isabelle, *Éclipses, comètes, autres phénomènes célestes et tremblements de terre au Moyen Âge. Enquête sur six siècles d'historiographie médiévale dans les limites de la Belgique actuelle* (Louvain-la-Neuve 1995).

Drake, Ellen Tan, *Restless genius. Robert Hooke and his earthly thougts* (New York and Oxford 1996).

Ducellier, 'Les tremblements de terre balkaniques au moyen âge: aspects matériels et mentaux', in: B. Bennassar ed., *Les catastrophes naturelles dans l'Europe médiévale et moderne* (Toulouse 1996) 61–76.

Ducos, Joëlle, *La météorologie en français au Moyen Age (XIIIe-XIVe siècles)* (Paris 1998).

Duffy, Christopher, *Siege warfare. The fortress in the early modern world 1494–1660* (London and Henley 1979).

Ehmer, Hermann, 'Zeichen und Wunder. Die theologische Deutung von Naturereignissen im nachreformatorischen Württemberg', in: *Blätter für Württembergische Kirchengeschichte* 88 (1988) 178–200.

Eisenstein, Elisabeth L., *The printing press as an agent of change* (Cambridge 1980).

Ellenberger, François, *Histoire de la géologie* (2 vol., Paris 1988–1994).

Esser, Thilo, 'Die Pest. Strafe Gottes oder Naturphenomen? Eine frömmigkeitsgeschichtliche Untersuchung zur Pesttraktaten des 15. Jahrhunderts', in: *Zeitschrift für Kirchengeschichte* 108 (1997) 32–57.

Ewinkel, Irene, *De monstris. Deutung und Funktion von Wundergeburten auf Flugblättern im Deutschland des 16. Jahrhunderts* (Tübingen 1995).

Feingold, Mordechai (ed.), *Jesuit science and the republic of letters* (Cambridge 2003).

Figliuolo, Bruno, *Il terremoto del 1456* (2 vol., Altavilla Silentina 1988).

Fiorentino, Silvia Grassi, '"Nella sera della Domenica...". Il terremoto del 1703 in Umbria: trauma e reintegrazione', in: *Quaderni Storici*, nuova serie 19 (1984) 137–154.

Frängsmyr, Tore, *Geologi och skapelsetro. Föreställninger om jordens historia från Hiärne till Bergman* (Stockholm 1969).

Fréchet, Julien, Mustapha Meghraoui, and Massimiliano Stucchi (ed.), *Historical seismology. Interdisciplinary studies of past and recent earthquakes* (Heidelberg 2008).

French, Roger K., 'Pliny and Renaissance medicine', in: Roger French and Frank Greenaway ed., *Science in the early Roman Empire: Pliny the elder, his sources and influence* (London and Sydney 1986) 252–281.

Freyberger, Pierre, 'Le problème du fatalisme astral dans la pensée protestante en pays germaniques', in: *Divination et controverse religieuse en France au XVIe siècle* (Paris 1987) 38–43.

Frijhoff, Willem, *Wegen van Evert Willemsz. Een Hollands weeskind op zoek naar zichzelf, 1607–1647* (Nijmegen 1995).

Gielen, Victor, *Das Eupener Land im Wandel der Zeit* (Eupen 1992).

Gilson, Etienne Henri, 'Météores cartésiens et météores scolastiques', in: idem, *Etudes de philosophie médiévale* (Strasbourg 1921) 247–286.

Gisler, Monika, 'Forschen in den "Eingeweiden der Erde". Johann Jakob Scheuchzers Erdbebenforschung zwischen Wissenschaft und Theologie', in: Simona Busoni Leoni ed., *Wissenschaft - Berge - Ideologien. Johann Jakob Scheuchzer (1672–1733) und die frühneuzeitliche Naturforschung* (Basel 2000) 73–87.

id., *Göttliche Natur? Formationen im Erdbebendiskurs der Schweiz des 18. Jahrhunderts* (Zürich 2007).

Gohau, Gabriel, *Les sciences de la terre aux XVIIe et XVIIIe siècles. Naissance de la géologie* (Paris 1990).

Graf, Fritz, 'Earthquakes and the gods: reflections on Graeco-Roman responses to catastrophic events', in: Jitse Dijkstra, Justin Kroesen and Yme Kuiper ed., *Myths, martyrs, and modernity. Studies in the history of religions in honour of Jan N. Bremmer* (Leiden and Boston 2010) 95–113.

Gragg, Larry, 'The Port Royal earthquake', in: *History Today* 50(9) (September 2009) 28–34.

Guidoboni, Emanuela, 'Riti di calamità: terremoti a Ferrara nel 1570–74', in: *Quaderni Storici*, new series 19(55/1) (1984) 107–135.

id., 'The oscillatory seismic motion and the daily motion of the earth in Francesco Travagini's *Physica disquisitio* (1669)', in: *Earth Sciences History* 37 (2018) 165–176.

Gulizia, Stefano, 'Cosmology and scholarship in seventeenth-century Helmstedt: the Baltic mathematician and scientific mediator Nicolaus Andreas Granius (c. 1569–1631)', in: P.D. Omodeo and V. Weiss ed., *Natural knowledge and Aristotelianism at early modern Protestant universities* (Wiesbaden 2019) 109–122.

Günther, Siegmund, 'Münchener Erdbeben- und Prodigienlitteratur in älterer Zeit', in: *Jahrbuch für Münchener Geschichte* 4 (1890) 233–256.

Gutdeutsch, Rudolf, Christa Hammel, Ingeborg Mayer and Karl Vocelka, *Erdbeben als historisches Ereignis. Die Rekonstruktion des Bebens von 1590 in Niederösterreich* (Berlin etc. 1987).

Hauer, Katrin, *Der plötzliche Tod. Bergstürze in Salzburg und Plurs kulturhistorisch betrachtet* (Vienna 2009).

Haynes, Alan, 'The English earthquake of 1580', in: *History Today* 29 (1979) 542–544.

Heitzmann, Christian, 'Gianozzo Manetti und das Erdbeben von 1456. Christlicher Humanismus und empirische Naturwissenschaft', in: A. Bihrer and E. Stein ed., *Nova de veteribus. Mittel- und neulateinische Studien für Paul Gerhard Schmidt* (Munich & Leipzig 2004) 735–748.

Hellmann, Gustav, 'Entwicklungsgeschichte des meteorologischen Lehrbuches', in: idem, *Beiträge zur Geschichte der Meteorologie*, II (Berlin 1917).

id., *Die Meteorologie in den deutschen Flugschriften und Flugblättern des XVI. Jahrhunderts* (Berlin 1921).

Hengst, Daniel den, 'The scientific digressions in Ammianus' Res gestae', in: idem, *Emperors and historiography. Collected essays on the literature of the Roman Empire*, D.W.P. Burgersdijk and J.A. van Waarden ed. (Leiden 2010) 236–247. (Reprint of an article from 1992.)

Heniger, Simon K., *A handbook of Renaissance meteorology* (Durham 1960).

Hoff, K.E.A. von, *Chronik der Erdbeben und Vulkanausbrüche*, I (Gotha 1840).

Hoffmann, Georg, *Sigismundus Suevus Freistadiensis. Ein schlesischer Pfarrer aus dem Reformations-Jahrhundert* (Breslau 1927).

Houtgast, Gerhard, *Aardbevingen in Nederland. Catalogus van aardbevingen t/m 1900* (De Bilt 1991) (KNMI-publikatie 1979).

Jakubowski-Tiessen, Manfred, *Sturmflut 1717. Die Bewältigung einer Naturkatastrophe in der frühen Neuzeit* (München 1992).

Jorink, Eric, *Reading the book of nature in the Dutch Golden Age, 1575–1715* (Leiden 2010).

Kempe, Michael, *Wissenschaft, Theologie und Aufklärung. Johann Jakob Scheuchzer (1672–1733) und die Sintfluttheorie* (Epfendorf 2003).

Kendrick, Thomas Downing, *The Lisbon earthquake* (London 1956).

Kennedy, John E. and William A.S. Sarjeant, '"Earthquakes in the air": the seismological theory of John Flamsteed (1693)', in: *Journal of the Royal Astronomical Society of Canada* 76 (1982) 213–223.

Kocher, Paul, *Science and religion in Elizabethan England* (San Marino 1953).

Kunz, Geert Georg and Dinant Petrus Oosterbaan, 'Buskruitramp was "straf of de zonden"', in: H.L. Houtzager et al. ed., *Kruit en krijg. Delft als bakermat van het Prins Maurits Laboratorium TNO* (Amsterdam 1988) 53–64.

Kusukawa, Sachiko, *The transformation of natural philosophy. The case of Philip Melanchthon* (Cambridge 1995).

Labrousse, Elisabeth, *L'entrée de Saturne au Lion. L'éclipse de soleil du 12 Août 1654* (The Hague 1974).

Ladero Quesada, Miguel-Angel, 'Earthquakes in the cities of Andalusia at the beginning of the modern age', in: Martin Körner ed., *Destruction and reconstruction of towns. Destruction by earthquakes, fire and water*, I (Bern, Stuttgart, Vienna 1999) 87–103.

Laureys, Marc, 'Die Bewertung der Prodigien und die Rezeption des Julius Obsequens im Humanismus des 16. Jahrhunderts', in: Wolfram Hogrebe ed., *Mantik. Profile prognostischen Wissens in Wissenschaft und Kultur* (Würzburg 2005) 201–221.

Lehner, Hans Christian, *Prophetie zwischen Eschatologie und Politik. Zur Rolle der Vorhersagbarkeit von Zukünftigem in der hochmittelalterlichen Historiografie* (Stuttgart 2015) (Historische Forschungen, 29).

Leppin, Volker, *Antichrist und jüngster Tag. Das Profil apokalyptischer Flugschriftenpublizistik im deutschen Luthertum 1548–1618* (Gütersloh 1999) (Quellen und Forschungen zur Reformationsgeschichte, 69).

Loeffler, Ulrich, *Lissabons Fall - Europas Schrecken. Die Deutung des Erdbebens von Lissabon im deutschsprachigen Protestantismus des 18. Jahrhunderts* (Berlin and New York 1999).

Ludwig, Walther, 'Pontani amatores: Camerarius und Hessus', in: Thomas Baier ed., *Pontano und Catull* (Tübingen 2003) 11–45.

id., 'Zukunftsvorstellungen in der Antike, der frühen Neuzeit und heute', in: Klaus Bergdolt and Walther Ludwig ed., *Zukunfsvoraussagen in der Renaissance* (Wiesbaden 2005) (Wolfenbütteler Abhandlungen zur Renaissanceforschung, 23) 9–64.

MacDougall, Doug, *Why geology matters. Decoding the past, anticipating the future* (Berkeley and Los Angeles 2011).

Martin, Craig, *Renaissance meteorology. Pomponazzi to Descartes* (Baltimore 2011).

Meinel, Christoph, 'Natur als moralische Anstalt. Die *Meteorologia philosophico-politica* des Franz Reinzer, s.j., ein naturwissenschaftliches Emblembuch aus dem Jahre 1698', in: *Nuncius* 2/1 (1987) 37–94.

Meier, Mischa, *Das andere Zeitalter Justinians. Kontingenzerfahrung und Kontingenzbewältigung im 6. Jahrhundert* (second edition, Göttingen 2004).

Methuen, Charlotte, 'The role of the heavens in the thought of Philip Melanchthon', in: *Journal of the History of Ideas* 57 (1996) 385–403.

Moroni, Andrea, and Massimiliano Stucchi, 'Materials for the investigation of the 1564, Maritime Alps earthquake', in: Massimiliano Stucchi ed., *Historical investigation of European earthquakes*, I (Milan 1993) 101–125.

Mosley, Adam, 'Past portents predict: cometary historiae and catalogues in the sixteenth and seventeenth centuries', in: Dario Tessicini and Patrick J. Boner ed., *Celestial novelties on the eve of the scientific revolution 1450–1630* (Florence 2013) 1–32.

Mouthaan, José, 'Early modern perceptions of natural disasters. The eruption of the Vesuvius in 1631', in: *Traverse* 10 (2003) 50–63.

Müller-Jahnke, Wolf-Dieter, 'Kaspar Peucers Stellung zur Magie', in: A. Buck ed., *Die okkulten Wissenschaften der Renaissance* (Wiesbaden 1992) 91–102.

Mulcahy, Matthew, 'The Port Royal earthquake and the world of wonders in seventeenth-century Jamaica', in: *Early American Studies* 6 (2008) 391–421.

Nassichuk, John, 'La description des "poutres" atmosphériques dans le *Meteororum liber* de Giovanni Pontano (v. 525–540)', in: T. Belleguic and A. Vasak ed., *Ordre et désordre du monde. Enquête sur les météores, de la Renaissance à l'âge moderne* (Paris 2013) 51–68.

Neilson, G., R.M.W. Musson, and P.W. Burton, 'The "London" earthquake of 1580, April 6', in: *Engineering Geology. An International Journal* 20 (1984) 113–141.

Nicolosi, Salvatore, *Apocalisse in Sicilia (Il terremoto del 1693)* (Catania 1982).

Oeser, Erhard, 'Historical earthquake theories from Aristotle to Kant', in: Rudolf Gut-deutsch, Gottfried Grünthal and Roger Musson ed., *Historical earthquakes in central Europe* (Vienna 1992) 11–31.

Oldroyd, David, *Thinking about the earth: a history of ideas in geology* (London 1996).

Osorio, Alejandra B., *Inventing Lima. Baroque modernity in Peru's south sea metropolis* (New York 2008).

[Overvoorde, Jacob Cornelis], 'Aardbeving 1692', *Jaarboekje voor Geschiedenis en Oudheidkunde van Leiden en Rijnland* 4 (1907) 47.

Pettegree, Andrew, *The invention of news. How the world came to know about itself* (New Haven and London 2014).

Poirier, Jean-Paul, 'Saints as protectors against earthquakes in popular culture in Italy and Latin America', in: *Earth Sciences History* 37 (2018) 157–164.

Prescher, Hans, and Otfried Wagenbreth, *Georgius Agricola - seine Zeit und ihre Spuren* (Leipzig and Stuttgart 1994).

Presser, Helmut, *Vom Berge verschlungen in Büchern bewahrt. Plurs, ein Pompeji des 17. Jahrhunderts im Bergell* (Bern 1957).

Ruggenthaler, Oliver, 'Erdbeben im Tiroler Inntal im 17. und 18. Jahrhundert', pdf on: http://franziskaner.members.cablelink.at/provinzarchiv/pa_erdbeben_tirol.pdf. (8 July 2016).

Quenet, Grégory, *Les tremblements de terre aux XVIIe et XVIII siècles. La naissance d'un risque* (Seyssel 2005).

Rauscher, Julius, 'Der Halleysche komet im Jahre 1531 und die Reformatoren', in: *Zeitschrift für Kirchengeschichte* 32 (1911) 259–276.

Rappaport, Rhoda, 'Hooke on earthquakes: lectures, strategy and audience', in: *British Journal for the History of Science* 19 (1986) 129–146.

Remmert, Volker R., '"Whether the stars are innumerable for us?" Astronomy and biblical exegesis in the Society of Jesus around 1600', in: Kevin Killeen and Peter J. Forshaw ed., *The word and the world. Biblical exegesis and early modern science* (New York 2007) 157–173.

Roebel, Martin, *Humanistische Medizin und Kryptocalvinisum. Leben und medizinisches Werk des Wittenberger Medizinprofessors Caspar Peucer (1525–1602)* (dissertation Heidelberg 2004).

Roger, Jacques, 'La théorie de la terre au XVIIIe siècle', in: *Revue d'Histoire des Sciences* 26 (1973) 23–48.

id., 'The Cartesian model and its role in eighteenth-century "theory of the earth"', in: Th. M. Lennon, J.M. Nicholas, and J.W. Davis ed., *Problems of Cartesianism* (Kingston and Montreal 1982) 95–112.

Rose, Paul Lawrence, *Bodin and the great God of nature. The moral and religious universe of a Judaiser* (Geneva 1980).

Rousseau, George S., 'The London earthquakes of 1750', in: *Cahiers d'Histoire Mondiale* 11 (1969) 436–451.

Rudwick, Martin J.S., *The meaning of fossils. Episodes in the history of palaeontology* (second edition, Chicago and London 1976).

Rueda Ramírez, Pedro, and Fernández Chaves, Manuel, 'El terremoto como noticia: relaciones de sucesos y otros textos del temblor de 1680', in: *Estudios sobre el mensaje periodístico* 14 (2008) 581–604.

Scaramellini, Guido, Günther Kahl, and Gian Primo Falappi (ed.), *La frana de Piuro del 1618. Storia e immagini di una rovina* (Piuro 1988).

Scarth, Alwyn, *Vulcan's fury. Man against the volcano* (New Haven and London 1999).

Schechner Genuth, Sarah, *Comets, popular culture, and the birth of modern cosmology* (Princeton 1997).

Schenk, Gerrit Jasper, 'ein Unstern bedroht Europa. Das Erdbeben von Neapel im Dezember 1456', in: idem ed., *Katastrophen vom Untergang Pompejis bis zum Klimawandel* (Ostfildern 2009) 67–80.

Schilling, Heinz, 'Job Fincel und die Zeichen der Endzeit', in: W. Brückner ed., *Volkserzählung und Reformation. Ein Handbuch zur Tradierung und Funktion von Erzählstoffen und Erzählliteratur in Protestantismus* (Berlin 1964) 325–392.

Schmidt, Peter, 'Deutschsprachiche Einblattdrucke seismologischen Inhalts des 16. und 17. Jahrhunderts', in: B. Fritscher and G. Brey ed., *Cosmographica et geographica. Festschrift für Heribert M. Nobis* (2 vol., Munich 1994) 387–395.

id., 'Seismologie und Medaillen vom 17. Jahrhundert bis in unsere Zeit', in: *Jahrbuch der Staatlichen Kunstsammlungen Dresden* 10 (1976–1977) 193–208.

id., 'Gedanken zum Umbruch in der europäischen Seismologie während des 18. Jahrhunderts', in: *Zeitschrift für Geologische Wissenschaften* 8 (1980) 189–206.

Schnurmann, Claudia, 'Das Erdbeben von Jamaica (Juni 1692) im zeitgenossischen Verständnis des englischen Kolonialreichs. Katastrophen als Mittel der Weltdeutung', in: Paul Münch ed., *"Erfahrung" als Kategorie der Frühneuzeitgeschichte* (München 2001) (Historische Zeitschrift, Beiheft (n.f.) 31) 249–259.

Schulze, Reinhard, 'Islamische Deutungen von Erdbeben und anderen Naturkatastrophen', in: Christian Pfister and Stephanie Summermatter ed., *Katastrophen und ihre Bewältigung. Perspektiven und Positionen* (Bern/Stuttgart/Wien 2004) 101–117.

Schwegler, Michaela, *"Erschröckliches Wunderzeichen" oder "natürliches Phenomen"? Frühneuzeitliche Wunderzeichenberichte aus der Sicht der Wissenschaft* (München 2002).

Scorsone, Antonio, *Giovanni Alfonso Borelli* (Palermo 1993).

Shapiro, Barbara J., *A culture of fact: England 1550–1720* (Ithaca 2000).

Sieberg, A, *Beiträge zur Erdbebenkatalog Deutschlands und angrenzender Gebiete für die Jahre 58 bis 1799* (Berlin 1940).

Sigurdsson, Haraldur, *Melting the earth. The history of ideas on volcanic eruptions* (New York and Oxford 1999).

Smoller, Laura, 'Of earthquakes, hail, frogs, and geography. Plague and the investigation of the Apocalypse in the later middle ages', in: Paul Freeman and Caroline Bynum ed., *Last things. Death and the apocalypse in the middle ages* (Philadelphia 2000) 156–187.

Solerti, Angelo, *Ferrara e la corte Estense nella seconda metà del secolo decimosesto* (Città di Castello 1891).

Stokes, Evelyn, 'Volcanic studies by members of the Royal Society of London 1665–1780', in: *Earth Science Journal* 5 (1971) 46–70.

Strake, Gerald M., *Anglican reactions to the revolution of 1688* (Madison 1962).

Streinz, Franz, *Die Singschule in Iglau und ihre Beziehungen zum allgemeinen deutschen Meistergesang* (Munich 1958).

Stucchi, Massimiliano ed., *Historical investigation of European earthquakes*, I (Milan 1993).

Sudhoff, Karl, *Bibliographia Paracelsica. Besprechung der unter Hohenheims Namen 1527–1893 erschienenen Druckschriften* (Berlin 1894; repr. Graz 1958).

Tahir, Mustafa Anwar (ed.), 'Traité de la fortification des demeures contre l'horreur des séismes (...), écrit à l'occasion du tremblement de terre de 984 H./ 1576', in: *Annales Islamologiques* 12 (1974) 131–159.

Taylor, John Gates, jr., *Eighteenth-century earthquake theories: a case-history investigation into the character of the study of the earth in the Enlightenment*. PhD-thesis, University of Oklahoma, 1975.

Taylor, Kenneth L., 'Before volcanoes became ordinary', in: W. Mayer, R.M. Clary, L.F. Azuela, T.S. Mota, and S. Wołkowicz ed., *History of geoscience: celebrating 50 years of INHIGEO* (London 2016) 117–126.

Tex, Emile den, *Een voorspel van de moderne vulkaankunde in West-Europa met nadruk op de Republiek der Verenigde Nederlanden* (Amsterdam 1998).

Udías, Augustín, 'Earthquakes as God's punishment in 17th- and 18th-century Spain', in: M. Kölbl-Ebert ed., *Geology and religion. A history of harmony and hostility* (London 2009) 41–48.

Van de Wetering, Maxine, 'Moralizing in Puritan natural science: mysteriousness in earthquake sermons', in: *Journal of the History of Ideas* 43 (1982) 417–438.

Vanpaemel, Geert, 'Comets, earthquakes and gunpowder. School philosophy in Libertus Fromondus' *Meteorologicorum libri sex* (1627)', in: *Lias* 41/1 (2014) 53–68.

Vermij, Rienk, 'Natuurgeweld geduid', in: *Feit & fictie. Tijdschrift voor de Geschiedenis van de Representatie* 3/1 (Fall 1996) 49–64.

id., 'Subterranean fire. Changing theories of the earth during the Renaissance', in: *Early Science and Medicine* 3 (1998) 323–347.

id., 'Erschütterung und Bewältigung. Erdbebenkatastrophen in der frühen Neuzeit', in: M. Jakubowski-Tiessen and H. Lehmann ed., *Um Himmels Willen. Religion in Katastrophenzeiten* (Göttingen 2003) 235–252.

id., 'A science of signs. Aristotelian meteorology in Reformation Germany', in: *Early Science and Medicine* 15 (2010) 648–674.

Vogt, Jean, 'Révision de la crise sismique rhénane de mai 1737', in: Massimiliano Stucchi ed., *Historical investigation of European earthquakes*, I (Milano: CNR Istituto di Ricerca sul Rischio Sismico 1993) 89–100.

Walker, Charles F., *Shaky colonialism. The 1746 earthquake-tsunami in Lima, Peru, and its long aftermath* (Durham and London 2008).

Weber, Christoph Daniel, *Vom Gottesgericht zur verhängnisvollen Natur. Darstellung und Bewältigung von Naturkatastrophen im 18. Jahrhundert* (Hamburg 2015).

Weichenhan, Michael, 'Astrologie und natürliche Mantik bei Caspar Peucer', in: S. Oehmig ed., *700 Jahre Wittenberg. Stadt Universität Reformation* (Weimar 1995) 213–224.

Weinberg, Joanna, '"The voice of God": Jewish and Christian responses to the Ferrara earthquake of November 1570', in: *Italian Studies* 46 (1991) 69–81.

Weltecke, Dorothea, 'Die Konjunktion der Planeten im September 1186. Zum Ursprung einer globalen Katastrophenangst', in: *Saeculum* 54 (2003) 179–212.

Willmoth, Frances, 'John Flamsteed's letter concerning the natural causes of earthquakes', in: *Annals of Science* 44(1987) 23–70.

Woitkowitz, Torsten, *Die Briefe von Joachim Camerarius d.Ä. an Christoph von Karlowitz bis zum Jahr 1553. Edition, Übersetzung und Kommentar* (Stuttgart 2003).

Zambelli, Paola (ed.), *Astrologi hallucinati. Stars and the end of the world in Luther's time* (Berlin 1986).

Zeller, Rosmarie, 'Wahrnehmung und Deutung von Naturkatastrophen in den Medien des 16. und 17. Jahrhunderts', in: Christian Pfister ed., *Am Tag danach. Zur Bewältigung von Naturkatastrophen in der Schweiz 1500–2000* (Bern, Stuttgart, Vienna 2002) 27–38.

id., 'Naturkatastrophen zwischen Kuriosität, Sensation und religiöser Interpretation. Zur Semiotik von Naturkatastrophen', in: Christian Pfister and Stephanie Summermatter ed., *Katastrophen und ihre Bewältigung. Perspektiven und Positionen* (Bern, Stuttgart, Vienna 2004) 79–100.

INDEX